SUPERYACHT SUCCESS

SUPERYACHT SUCCESS

Your guide to navigating the yachting industry

BRENDAN O'SHANNASSY

ADLARD COLES

LONDON • OXFORD • NEW YORK • NEW DELHI • SYDNEY

ADLARD COLES
Bloomsbury Publishing Plc
50 Bedford Square, London, WC1B 3DP, UK
Bloomsbury Publishing Ireland Limited,
29 Earlsfort Terrace, Dublin 2, D02 AY28, Ireland

BLOOMSBURY, ADLARD COLES and the Adlard Coles logo are trademarks of
Bloomsbury Publishing Plc

First published in Great Britain 2025

A catalogue record for this book is available from the British Library

Library of Congress Cataloguing-in-Publication data has been applied for

ISBN: HB: 978-1-3994-2649-7; ePub: 978-1-3994-2647-3; ePDF: 978-1-3994-2648-0

2 4 6 8 10 9 7 5 3 1

Typeset in Sabon LT Std by Deanta Global Publishing Services, Chennai, India
Printed and bound in Great Britain by Clays Ltd, Elcograf S.p.A.

MIX
Paper | Supporting
responsible forestry
FSC® C018072

To find out more about our authors and books visit www.bloomsbury.com and sign up
for our newsletters

For product safety related questions contact productsafety@bloomsbury.com

CONTENTS

FOREWORD

Throughout every story, where I have a walk-on role, I have sought to be truthful to my memory and represent my emotions at the time. Given that the arc exceeds 35 years I am aware that my recall has been sliced by the ravages of time: sometimes a story is retold so many times it becomes the driver and I feel the passenger to the tale. I hope this is not the case on too many occasions, but I accept my memory of a day 20 years past may vary from that of the person I was sitting next to at the time.

As with my first book, *Superyacht Captain*, this book was never intended to be a 'kiss-and-tell' of the yachting community and those that play their roles within it. Yes, an informed reader may recognise details in a tale and may even be able to identify what and who is involved. To avoid this entirely would be to shield the truth and I could not do this. Where I think there might be any negativity, I have sought to place enough of a veil to obscure the bride. Some names are changed, some are not, and others just not mentioned.

As much as I enjoy writing, I love reading more. I do not conceal that many insights are my repackaging of the thinking of those who have walked ahead of me. I chanced upon a wonderful quote from Shane Parrish of Farnam Street that describes my aim: '[to] master the best of what other people have already figured out.' I don't claim any greater insight than what my experiences and my reading have given me.

INTRODUCTION

A carrier wave is the frequency that alone does not transmit information: the information is overlaid through modulation, either by amplitude (AM) or frequency (FM). I learned this enjoyable fact during my naval undergraduate days when we were studying frequency propagation, and the concept seemed so elegant in its simplicity. I was somewhere between 17 and 18 years old and until that point the radio or television had to my mind just been a stand-alone unit; sure, I knew that TVs were receivers and I knew the terms AM and FM well enough, but I'd never put it together that the static between stations was the noise of the frequency without any information being carried.

Opening a book with a description of the absolute basics of the radio spectrum might seem somewhat esoteric, but the concept of a carrier wave to propagate a message has stayed with me. This book explores and uses the maritime environment for its message, the yachts and experiences are my carrier waves, and the learning outcomes are the modulation.

You might ask why the marine environment is a good frequency to be used as this carrier and, more specifically, why my own career, largely spent as a captain on superyachts, is worth sharing? Well, captaincy provides in real time a place where leadership and operational competency are tested in parallel; like operating theatres or plane flightdecks, the bridge is a high-stakes, high-pressure environment that accelerates learning and leadership. It's an arena where success and failure sit constantly on a knife edge and where the feedback loop on performance is as close to instantaneous as is possible. There are many examples I could give, but some that are most identifiable are running a ship aground, missing a turn, misreading a chart – all of which can turn an otherwise pleasant day into a life-and-death situation. To be clear, I have never grounded a ship, but I have seen fires, life-threatening accidents, physical and emotional collapse of crew, and weather conditions that threatened the vessel and all on board.

Added to this constant, unrelenting strain of being exposed to risk, and in contrast to almost all other workplaces, a captain does not leave the team at the end of the day's work. Indeed, the day does not end when you are at sea. The surgeon, the pilot, the sports star and the CEO all get to go home at the end of their day. The captain is tested, 24 hours a day, for weeks and maybe months at a time, leading tens, hundreds and in the case of cruise ships, thousands of crew from diverse cultures and backgrounds. The marine environment does not know the day of the week, the time of day, whether you are tired, awake, asleep or sitting on the toilet. I mention the last as I have lost count of the number of alarms that have been activated when I have gone to the toilet.

My first book, *Superyacht Captain: Life and leadership in the world's most incredible industry* shared my experiences and learnings from captaincy. I now take a deeper dive into what was going on as these experiences unfolded. I examine the human perspective of an industry that is often defined just by the yachts and the glamorous guests. This is not uncommon – public companies are measured by share price movement and construction firms are judged by what they build – but nothing happens without leaders at the helm. I know in the case of yachting, the glamour provides the momentum for the media to take an interest in the industry, though in the same way that grand houses would be nothing without the downstairs staff (a gentle nod to *Downton Abbey*), yachts cannot function without the crew.

Along with using the yachts to carry a message, I will provide an insider's guide to how to succeed in the normally shielded world of superyachts, both on board and at sea. I am able to do this thanks to more than two decades of sitting 'inside the room' of the industry. I am not a features reporter or someone who has stepped in and out of the industry. I have made a life's career within the superyacht community. Indeed, much of this book was written while I was on board a 90-metre yacht berthed in Monaco. As I write today, it is a week before the annual Monaco Yacht Show – the epicentre of a global industry, my experiences of which I will share with you, starting with the very first yacht on record, through to my own first day and finally to where I sit now.

As is my way, there will be diversions. My stories are woven around the yachts, the guests, the crew and the locations; these are the stage and cast for the bigger messages of the book. Stay with me as I take long walks through yachting stories; if I was a turnip farmer, they would be turnip stories, but yachting is what I must work with. I guarantee that within each tale there will be something that you recognise, since I believe that our thoughts transcend the environments that create them. This book offers a chance to learn from a high-performance environment that could equally benefit someone in sales, service, operations, construction, finance and almost every sector in between.

Anecdotal divergences aside, there is a structure to the book and the chapters within it. In Chapter 1, we will investigate life before yachts, to set the scene and give you a little insight into this new world. Next, we leave the dock and begin a life spent floating and living on board. After some thoughts of how to begin in the industry, in Chapter 2 I introduce you to the yachting community via some of the history and details of the superyacht diaspora, including their language, to give you a little taste of what the industry is about.

Chapter 3 moves on from the broad-strokes background to the most important members of the yachting community – those who choose to invest their funds and time in the creation of crazy and audacious yachts for no other purpose than recreation: the owners.

If the owners are the most important people in this tale, then ranking just behind them are the other humans on board a superyacht: the crew. For anyone wanting to understand team dynamics, Chapter 5 is the place to go, since it examines how these play out on a superyacht. This is the most personal chapter for me: writing about my favourite yachts or professional competencies does not expose my human frailty, whereas deconstructing personal interactions past and present does. It is the area of my own career where I have had the most growth and is where I have the most work to do in the future.

After describing all the people involved in the yachting circus, the book moves on to describe the actual job on board. As with so many workplaces, there are many skills required, not all of which can be covered in one book. The chapters in this section instead comprise stories from my own career, with many universal takeaways and observations that will apply whether you are the captain, the deckhand, have other roles on board, or are not even at sea. Topics covered range from essential skills for new crew on board, through interpersonal skills that will help you cope with the role and lifestyle, to how to be a part of a happy crew.

The final section is the most how-to part of the book, providing four toolboxes for making a success of a life at sea, whether that's communicating effectively, getting along with your peers, cultivating crew lifetime value or improving your core competencies.

However, having laid out the broad skeleton of the book, please note that it can also be dipped into in a random order and in small amounts. If it were a meal, think more tapas than three-course formal dining. Since I am truly guilty of stretching metaphors beyond their use-by date, I am going to ask that in reading you do keep in mind the carrier wave concept I opened with. Stay with me, even if the story seems to be a long way from your own experience or what you seek to develop. Trust me: there will be a message – the modulation – in there. Again drawing upon the radio frequency spectrum for guidance, hopefully the signal-to-noise ratio is high.

If you want to join me and carve out your own career in this challenging, exhilarating and dynamic industry, then this book is for you. You will gain insight, a new language, and a sense of where the dangerous waters are and how to avoid them. Of course, you may already be well advanced in your knowledge and experience in yachting, and may disagree with my observations. That's your prerogative. I merely provide the information, observations and anecdotes based on several decades' worth of experience in the hope that they give pause for reflection. If you were fortunate to have read my first book, then I open as it closed. If you wish to be a critic, sit down, start writing and don't stop until you've finished. The pain goes with being in the arena, not sitting on the sidelines. So without further digressions, let's get going!

PART ONE

INTRODUCING THE INDUSTRY

In this first section we have a chance to begin to frame our superyacht life. What does it look like from the outside? What will those first conversations among yacht crew feel like? And what is the language of this new world? These may seem quite simple, but when setting your sights on success, it is advisable to take every little advantage that is offered. That is my guarded way of saying, 'Don't skip this little introduction, as it might contain just the thing you need to stay metaphorically afloat in the early days.'

I

LIFE BEFORE SUPERYACHTS

I've always been drawn to boats. Having grown up in coastal Western Australia, I felt more at home on the water than in the school bus. By my mid-teens, I was buying the expensive boating magazines and daydreaming of yachts in some yet to be defined form. This was a long time before I finally boarded a long-haul flight from Australia to Nice, France, in 2001. I was risk-averse and only finally made the 'great leap' because my childhood friend Nick held a place for me on the yacht he was working in. For the longest time I had looked upon Nick with healthy envy when he would return to Australia with seemingly more than ample funds to spare, showing me photos and telling stories of the very places and yachts I was paying to look at in magazines. My hesitation to make the move myself spoke to a lack of confidence and a lack of worldly perspective: I may have held court in Sydney about the harbour sailing scene (well, I would have if anyone would have listened), but until Nick threw me a lifeline, I lacked the confidence to put that all in a bag and actually show up on the other side of the world and give it a go.

There is a hint of irony that today whenever I am approached by would-be yacht crew – often the offspring of friends, friends of friends or absolute random connections – I tell them they must jump the gap first. A common thread I've noticed in many of these conversations is that they are looking to get their start without leaving home. Maybe I am compensating for my own lack of gumption, but my guidance is that they need to get out of their postcode and go where the yachts and docks are, with 'shoulders back and smiles wide'.

I am writing some of this book on board in Monaco, where I can see young hopefuls walking along the docks, handing over CVs and hoping for their chance. I view them with deep respect: they are doing something I did not have the courage to do, and due to this I always give them my time. The best of them have bought and wear the clothes that mirror crew uniform; they look like they belong. The very best ask me my name, hold my attention, and tell me what they could deliver to the yacht if I were to give them a chance. If I could, I would hire them all.

KEY TAKEAWAYS

- Yachts travel globally, but hire locally, so the very first step on the journey towards superyacht success is to travel to where the yachts are and put yourself in the wind. At the very worst, you will have a holiday in a great location and realise that a big yachting adventure might not be for you. At best, you might start out on a new course.
- Present yourself smartly and wear clothes that mirror crew uniforms so that you look like you belong.

LIFE ASHORE/LIFE BEFORE

Talking within crew, a common opening to a conversation will be 'What did you do before yachting?' The fact that this is such a common question acknowledges both the diversity of humanity drawn to the industry and the fact that, not unlike the military, yachting will park your history, put you in a uniform, strip you down and then build you back in its image.

In the early days of your career, your side of a conversation will likely mostly consist of stories of home. This may feel awkward, as your new colleagues regale you and each other with tales of their favourite places in the Pacific, Caribbean or Indian Oceans. Your own versions of these experiences lie in your future, and that is OK. The blasé way that crew will speak of their favourite bars in three continents can seem borderline offensive to the wider world that does not travel at whim, but on board, 'normal' will shift, and this leads to some unexpected outcomes.

This book will not hide the fact that some people lose their way in the wonderful world of yachting, but by keeping your links, your connections and your imprinting to home, you can help yourself maintain your sense of self that yachting often seeks to tear away. I rely on my family and closest friends to keep me tethered. However, I do not profess to be blemish-free, and in this book I will share where and how I have slipped up. I may not explicitly state it, but an aware reader will be able to spot when I let the juggernaut of a superyacht take me over.

I will come back to my story later, but before I do I need to take a step back and look at the industry from a distance, taking a meander across history, psychology and the yachting vernacular to provide some framework and reference points for the later suggestions for how to succeed as a son or daughter of yachting.

KEY TAKEAWAYS

- Don't let the fact that you don't have yachting stories faze you in the early days.
- Maintaining links with friends and family outside of the yachting community is very important.

THE LANGUAGE OF YACHTING

The name of this section includes the words 'the industry'. To clarify, by that I mean the industry around superyachts. Although outsiders might talk of 'megayachts', 'gigayachts', 'white boats' or even the universal term 'superyachts', when professional yachting peers are talking about the profession, we are more comfortable using the terms 'the yachting community', 'the yachting industry' or really, simplifying it to just 'the Industry'.

There are many other such insider terms and phrases, and it pays to take the time to learn them before you start trying to get your first job. There is the obvious nautical parlance, such as names for parts of the ship and crew titles, which are essential knowledge, but you'll find your first few weeks much easier if you are familiar with some of the more esoteric language too. The best place to learn this in advance is to listen: listen in your reading, and when you begin your search, listen to those just a few paces ahead of you in the journey.

You also shouldn't be afraid to ask if you're unsure what something means once you are on board. It's much better to seek an explanation than to pretend to understand when you haven't. A moment's awkwardness is far preferrable to an extended period in ignorance.

KEY TAKEAWAYS

- Learn the language before you start, but do not overplay your hand once you start your job aboard.
- For more experienced crew, keep the jargon light but correct for the listener and be prepared to explain what something means to those who are newer to the industry – be the enabler, not the barrier.

2

THE HISTORY OF YACHTS AND YACHTING

To acknowledge that your own time at the helm of your chosen path is merely part of a continuum is to become a student of history. From fast food and fashion to finance, there is always a foundation story to be learned. In the case of yachting, it is one peppered with adventure, misadventure, happiness and the tragedy of war. The most wonderful new yacht of today is an evolution of all that has gone before it.

Given our species' longstanding attraction to the sea, it is hard to define when 'yachting' as a pastime really began. It is human nature that as soon as two boats are on the water one will seek to race the other and in doing so derive a pleasure beyond their original purpose, be that fishing, warfighting, exploring or trading (pillaging) on behalf of the sovereign.

FROM *MARY* TO *SUNBEAM*

Yachting today is as varied as the designers can imagine and the builders can build. This modern diversity somewhat obscures the industry's common ancestry, though the name hints to a genesis, a point of beginning. The Dutch *jachts* were the light, fast sailing vessels used to pursue pirates and privateers in the coastal waters of Western Europe. The Dutch then took their *jacht* concept from military to symbolic when Her Majesty's Yacht *Mary* was gifted to the English monarch Queen Mary by the Dutch East India company in 1660 to serve as a pleasure craft. Today, the definition of a yacht is a craft that can be powered by motor or sail and that is used for pleasure or racing.

If this gift to Queen Mary is seen as the 'Big Bang' moment in yacht history, then it provided a troubled start: HMY *Mary* sank in fog after only 15 years and the Dutch went to war with the British twice more in living memory of the gift. Even with this less than auspicious start, yachting quickly became a dominant economic force in England, growing from privateers to the Royal Navy and the great tea clippers.

'In 1846, Britain boasted 530 yachts for a total of 25,000 tonnes and employing 4700 men [...] by 1884, the number had increased to 2300 yachts...'

When I was starting out, I took the time and effort to learn the history of yachting. There is a reason why nations, militaries, successful firms and great families all uphold and celebrate their histories: it is the glue that binds them to a common goal, defines their cultures and guides their futures. I could not have articulated this when as an 18-year-old undergraduate I was studying naval history and had to learn about the great battles and leaders of the very young Royal Australian Navy. (Heaven forbid I had been born in China, where there are thousands of years of nautical history to learn!)

I confess that at the time I saw the learning as a distraction. Surely I should be studying everything out in front and not wasting time looking backwards? I was very wrong. If the future is the wind in the sails pushing forwards, then history is the keel, slowing the movement down a little, but balancing the force.

Despite my reluctance to study naval history in – it was a different time and I held different motivations – I was enjoying learning about the history of yachting and was keen to find a story that described the beginnings of this relatively modern genre. There are many potential choices, but I chose the sailing yacht *Sunbeam* as my reference to mark the beginning of modern global yachting. Launched in 1874, *Sunbeam* was a fully crewed vessel that allowed a family of means to complete a global tour. Anne Brassey's subsequent narrative of the journey, titled *A Voyage in the 'Sunbeam': Our Home on the Ocean for Eleven Months*, is fabulous. The book remains in print and is well worth a read – there is much more to this story than I've described here, such as Anne Brassey's influence on the modern Red Cross. (A common thread in yachting history, and this book, is that where there is great adventurous wealth, there is often great philanthropic commitment.) I am not going to overplay this tune and try to place yachts as the unacknowledged conduit of humanity's greatest leaps forward, but nor will I allow their demonisation. This is something we will come back to later.

BRASSEY'S YACHT, *SUNBEAM*, UNDER FULL SAIL
(COURTESY OF STATE LIBRARY VICTORIA/WIKIMEDIA COMMONS)

Racing forward from *Sunbeam* consciously omits some incredible yachts. The history of yachting is sufficient to fill many books and many coffee tables alone. A pause and a nod are needed for the period around the 1930s, particularly with several yachts from this era remaining in service and being without peer for their classic cachet. These include *Shemara*

CLOCKWISE FROM TOP LEFT: SHEMARA (COURTESY OF Y.CO), ARGOSY (COURTESY OF CLASSIC YACHT ASSOCIATION), DANNEBROG (COURTESY OF YACHT CHARTER FLEET), SAVARONA (COURTESY OF YACHT HARBOUR) AND HAIDA (COURTESY OF EDMISTON)

(1938), *Argosy* (1931), *Haida* (1929), *Dannebrog* (1931), *Nahlin* (1930), *Talitha* (1929) and *Savarona* (1931), all of which remain classics. The glorious *Christina O* (1943) also has its mythical place in the genealogy of yachting. These yachts are substantial, even by today's metrics, and their beautiful lines have held through the decades. In common with businesses spanning the full 20th century, these yachts have lived through times of great extravagance and hardship and then periods of secondment to various nations' war efforts. As a case in point, *Christina O* began as a Canadian warship, was left to decay and then bought and turned into 'Camelot afloat' by the business magnate Aristotle Onassis. I have passed by this yacht so many times over the years and every time, I think of the legacy that all on board must carry.

With yet another enormous leap I look to the yachts of the 1970s – a time when the world was changing, and Studio 54 and disco were breaking cultural taboos. The yachts and those who designed them were not going to be left behind and were breaking boundaries too, including one man who would go on to redefine yachting: John Bannenberg.

John Bannenberg was a maverick who turned the design of yachts on its head. When in 1971 *Corinthia VI* was launched, a new era of yachting had arrived. The owner, Austrian supermarket magnate Helmut Horten, had commissioned a yacht that was to be fast and 'warship like', and Bannenberg turned this brief into the defining yacht of the generation. This was accompanied by equally challenging designs, including *The Highlander*, *Nabila*, *Azteca*, *Paraiso*, *Coral Island* and, in homage to *Corinthia VI*, *Limitless*. I sit alongside many in the industry when without a missing a beat I speak of *Corinthia VI*, as the greatest yacht of all time.

A passing side note to *Corinthia VI*: the yacht was built as *Corinthia V*, but after hitting an uncharted rock (they always are), it sank on its maiden voyage in the Mediterranean. The owner immediately ordered a replacement, *Corinthia VI*. Wisely, this had better watertight subdivision than its ill-fated predecessor.

Corinthia VI (Courtesy of Superyacht Times)

The Bannenberg yachts did not sit as outliers in a design field of their own: they were a catalyst spurring other designers to innovate. From Bannenberg's work, the boundaries had been redrawn and the designers moved out to the new edges. Martin Francis, Tim Heywood, Ken Freivokh, Terence Disdale, Andrew Winch and Espen Øino would all go on to take yachting to new heights.

As the story of this book closes in on the present, the legacy of these yachts cannot be forgotten. They represent a history of passion, a history of power, a history of excitement. The yachts, their owners and their operating magnitude send a shiver, a sense of true awe, down yacht lovers' spines. There is also a human impact from the swirl of a designer's pen, for they move the market to larger or smaller vessels and create the environments that will shape the careers of thousands who will flow on and around and in their designs. As with many industries, there are always great innovators who change process forever, and even when their innovations are overtaken by technology and progress, their influence remains.

I cannot finish this whirlwind trip through yachting history without going backwards in history to a person who sadly is known more for his death than his life: Archduke Franz Ferdinand of Austria, whose assassination by the Serbian Gavrilo

Princip is correctly cited as one of the major contributing factors in the outbreak of the First World War and the subsequent destabilisation of the world for much of the 20th century. What is not as well recorded was that the archduke undertook one of the most audacious sovereign voyages in history. Beginning in 1892 and continuing throughout 1893, the archduke sailed under the comical pseudonym 'Count Hohenberg' on the torpedo ram cruiser *Kaiserin Elisabeth*.

The voyage began in Trieste (then Austrian) and continued as a circumnavigation through the Suez Canal to Egypt, India, Australia, Indonesia, New Guinea, Borneo, China, Japan, Canada, the United States and back to Europe, ending again in Austria (today's Italy). The stated goal was to find warm weather to improve his health, but alongside this was much hunting and general entertaining – a yacht cruise for the ages. It would be fair to say that the archduke would have been one of the most travelled humans on the planet at the time and in his diaries there is a sense of awe and humility in his depictions of the things, people and places he encountered while cruising.

It could be a stretch, after all Franz Ferdinand was a Habsburg archduke, but I like to think that perhaps if he had not been assassinated his global perspective and awareness of the lives of others could have taken the 20th century in a very different direction. Revisionist histories are the stuff of fantasy but think of a 20th century where the horrors of the two world wars were replaced with leadership by a monarch committed to bringing the world together and not tearing it down. One can but wonder what that would have been like.

WE NEED A BIGGER BOAT

I remain enamoured with nearly everything that floats. I look as longingly at a well-proportioned and appointed fuel barge in Singapore as I do at a 5-metre sailing sloop on an Austrian lake. There is an innate beauty in well-proportioned craft, and particularly in sail, it is rare for an efficient hull form and sail plan not to hold a beauty. Extending this to modern superyachts, it would be a shorter story to name the few yachts that I do *not* see beauty within. As this would be mean-spirited, indulge me as I stick with my longer list of 'great yachts' – their greatness based on their design, technology, impact on the yachting community or use.

Even with a nod to beauty and innovation I see many pathways for yachting 'greatness' and acknowledge that everyone in the yachting industry will have their own list, which will likely vary considerably from mine. As the introduction announced, they can share it when they write their book. My own superyacht journey is far from complete, and I expect my list to grow as I do. In addition to the listed criteria some of the 'great yachts' are included due to my personal association and others – well, I hope it is obvious – are here because they are wonderful.

Nabila (ex-Trump Princess; current Kingdom 5kr)

Launched in 1980, *Nabila* stole the show for me when a young Kim Basinger appeared on deck as Domino in the 1983 Bond film *Never Say Never Again*. Given it was owned by a real-life arms dealer, it was a completely appropriate lair for Bond villain Blofeld – sometimes art imitates art. It had lines that didn't seem to work, but somehow still did. It reminds me now of a floating DeLorean car. It is an absolute outlier on my list – a 'counter-classic' – since I do not know very much about the yacht but, like the DeLorean, even though I don't know if I even like it, I cannot look away.

Kingdom 5kr now spends much of the year on its home berth on Quai des Milliardaires in France, otherwise known as International Yacht Club Antibes (IYCA). With its hard edges it looks out of place among the newer yachts with their curves. I enjoy watching younger yacht crew being confronted by this and commenting 'What is this thing sitting here?' I can but smile and acknowledge that all beauty queens age, and fashion changes with the generations.

Lady Moura

Lady Moura was quite new when I began yachting and it simply overwhelmed me. I could not fathom that such a yacht could be in private ownership. I walked behind it on the dock in Monaco, I walked alongside it, and I looked towards it as I washed the foredeck of another yacht nearby. Years passed and yachts became bigger, but *Lady Moura* retained its place among my list of greats.

There are stories based in fact and stories based in legend that swirl around *Lady Moura*. I do not know which is which, and nor would I want to. As my career developed, I became the peer of captains I had once been in awe of. *Lady Moura*'s captain was one of these and I was like a giddy schoolboy when Matthias invited me for a tour of the great lady. It is a dangerous thing to meet one's idol and I knew that seeing *Lady Moura* up close and personal was a risk. My eye now being that of a superyacht captain trained to see the details and the flaws, I am no longer distracted by the scale and majesty of a yacht. Yet even with this perspective the great *Lady Moura* did not disappoint. The years date some of the finer points of the interior design, but I left the tour feeling even more enamoured with the operation than I had been from a distance.

Octopus

I remember the first time I saw *Octopus* as I approached driving a tender from a 63-metre yacht also anchored in Saint-Jean-Cap-Ferrat. I could not believe what I was seeing; the transom alone was so daunting in comparison to the 5-metre tender I was driving. Not unlike *Lady Moura*, there are stories upon stories that surround *Octopus*. Fortunately, having been one the few captains of this fabulous yacht, I know the truth and think the facts are more impressive than any fiction could be. The yacht was commissioned by Paul Allen, co-founder of Microsoft and a visionary yacht owner. This vision allowed a construction team to go further than anyone in yachting had previously attempted: the build team needed to utilise commercial and military technologies to deliver the crazy requirement to launch a 20-tonne, ten-person submarine and a 23-metre cabin cruiser from a wet dock within the yacht. The resulting 'marina' included a 90-tonne hinged plate in the hull that would increase the yacht's draught from 6 to 11 metres. The hangar above the marina could accommodate two helicopters, with the larger being a Sikorsky S-76. The hydrographic installation could map the seabed and simultaneously track the submersible and a remote-operated vehicle, the latter with the ability to descend to 3,000 metres.

The yacht was built for polar exploration before the term 'explorer yacht' entered the lexicon of yachting. During my induction tours of *Octopus*, I would struggle to maintain a professional nonchalance as I was shown yet another feature. Unlike the crew who were experienced on the vessel and took some of it for granted, for me, every corner unveiled some treat. Maybe it was the ice-class glass-bottomed viewing area, possibly the six-deck guitar string art installation or simply the hidden cranes that allowed stores to be loaded. It all spoke to a clarity of vision absent in so much of our daily lives.

Motor Yacht A (MY A/Sailing Yacht A)

I did not mention Philippe Starck among my list of 20th-century yacht designers simply because of his ubiquity of design. However, no modern story of yachting can omit the A yachts. Personally, I fell in love with *MY A* at first sight. I had been away from the Mediterranean for some time and had seen the photos and heard the stories, but it was not until in falling light I anchored adjacent to the great yacht in Antibes that I fully appreciated the achievement. In one construction, yachting had begun a new epoch. It was wrong, but it was also just so right. It also raised a few questions: could it submerge like a fantasy ship? Where and how do the people even get out on deck?

Unlike my immediate attraction to *Motor Yacht A*, I struggled with its younger and bigger sibling, *Sailing Yacht A*. I was

16

working in North Germany during the sail yacht's construction in Kiel, and I would often drive past the Nobiskrug shipyard, where I'd snatch a glimpse of the hull form or, after a time, the three towering 100-metre masts that dominated the skyline. It wasn't working for me: there was something about the slope of the sheer line and the forward curve of the masts that offended my sense of 'how it should be'. Time passed. I saw the yacht at sea and gazed upon the professional photographs that began to circulate – all in the best light and showing the most flattering angles. Still, I was not swayed: *Sailing Yacht A* was not working for me.

Until, that is, one glass-calm day off the coast of Argentario, Italy. I was captain of another yacht of comparable size, and we were shadowing the great sailing yacht along the coast. The two bridges were in contact by VHF radio and with *Sailing Yacht A*'s blessing we closed much closer than would normally be courteous. As we did so, something began to happen. I started to see the details: an oval window normally hidden from view, a children's play area visible from the stern, and a folding terrace being lowered ahead of the yacht coming to anchor. In a way that mirrors life, it is not the 'love at first sight' moment that has the depth, it is learning the intricacies of another and falling deeper and deeper with each observation.

I raised the binoculars to observe the detail more closely and then I saw a crew member walking by the bridge deck. Their size was the first reference I had: they looked tiny. The two yachts anchored in the sheltered waters just north of Punta Ala on the Tuscan coast. For once it was a peaceful afternoon and when the captain invited me for a tour I leapt at the chance. Being able to visit and see this commitment to design excellence first-hand remains a career highlight. It is not for me to puncture the mystique of this grand yacht with a poor description of the designer's vision, so suffice it to say that the closer I looked, the deeper in love I fell with it. There are naysayers who speak of its lack of practicality. To those I say, yes, it is entirely unpractical, but then beauty is not always about function.

Matching the mythology of this great yacht was its captain. Claus and I had known each other for 20 years, but as is so often the case in yachting we had only met in person once or twice. Hailing from a multigenerational German seafaring family, Claus embodies all that a yacht captain could and should be. Walking through *Sailing Yacht A,* Claus's manner was as striking as the unbelievable achievement of the concept, design and construction. He was relaxed and convivial – we laughed as we walked – but his eyes gave him away: he was constantly glancing to every corner, to every cushion, to every line, to every engine part in the storeroom. As we walked and shared our stories, I knew Claus was building a catalogue that would flow down to the department heads by way of gentle admonishment and a call to raise their own standards to mirror his and those the yacht's design demands.

'If you have to raise your standards to meet my minimum, this is not going to work.'

Somewhere during our tour Claus shared one of the comments he uses during recruitment: 'If you have to raise your standards to meet my minimum, this is not going to work.'

This phrase sums him up. These are not only words: they are how Claus lives his life. He is not stiff or unapproachable – there is a gleam in his eye and a smile for all the crew. His approach clearly works. It was very discreet, but I noted during my visit that each crew member instinctively adjusted some aspect of their dress and bearing before returning Claus's smile with their own. He is a great captain matched to a great yacht.

Ulysses/Andromeda

Fundamentally, there is nothing in *Ulysses* (now *Metaverse*) or *Andromeda* that means that they should make this list. They are not elegant, beautiful or even built to the highest standard. In stark contrast to the previous entries, these yachts are a triumph of function over form. They were made by a serial yacht builder who did not care for established norms, and the true reason they made this list is that *Andromeda* and later *Ulysses* shook the cage of the large yacht community simply by making designers and builders look more closely at what they were offering.

I was not so sure I agreed with the function over form notion to begin with. My passion for yachting beauty and vanity clouded my commercial view of practicality. This view was laid bare when in discussion with a yacht owner he reviewed the value proposition of *Ulysses* impartially – not unlike the way he would review something in his day job as a venture capitalist undertaking due diligence prior to any investment. This potential yacht owner listed *Ulysses'* capabilities, compared these with the competitors in the market at similar cost, and found a definitive conclusion: *Ulysses* was the better option. The metrics were simple but could have been missed by a more aesthetically focused yacht owner. I let that sink in. With only some reduction in aesthetic reward, you can have a more useable yacht for half the price.

Economic factors aside, there is another reason these yachts made my list: I served as captain for two years with them and saw their capability for myself. If beauty also means the ability to get up each day and do the job, then these two workhorses fully deserve their place here.

Mirabella V/Maltese Falcon/Black Pearl

Much of this book leans to power ahead of sail, but the sailing beauties *Mirabella V*, *Maltese Falcon* and *Black Pearl* cannot be left from this list and the focus on power reflects my experience rather than being a preference. I am the first to acknowledge the beauty of a large sailing yacht and marvel at the engineering in place to handle the sails without compromising the aesthetic, but while sail may hold my heart, it is not where I have made my career. There was an inflection point within my career where I may have turned to sail; it was a road that diverged in a yellow wood and I can no longer look to what might have been, though I can still acknowledge amazing sail yachts.

Mirabella V was reported on eagerly when it was launched and as the number in the name suggests, it represents a yacht owner's evolution. Joe Vittoria was a serial yacht owner, and the *Mirabella* brand was strong by 2004 when *Mirabella V* was beginning to cruise. I recall an article that described the challenges of building it, including the fact that the 88-metre mast was at a different scale to anything else around at the time. There was a quote from a crew member that spoke about everything normalising for them after a while and then something would happen and they would say to each other, 'It's doing that big thing again.' This caption was set against a photo of the stainless-steel forestay turnbuckle that at 2 metres towered well above the crew member. My experience of foremast turnbuckles was picking one off the shelf at the yacht chandlers.

At the time of *Mirabella V*'s launch my years on sailboats remained at the forefront of my memory and I scoured the yachting anchorages eagerly for my first sighting. It came on a quintessentially calm morning in Portofino. Heading south just after dawn, I rounded the special navigational buoy that borders the Portofino National Park and began looking to the radar for a suitable anchorage. The automated information that all vessels transmit showed *Mirabella V* adjacent to my planned anchorage before I could see her by eye. My heart quickened; finally, I would see this incredible Ron Holland-designed, British-built testament to design and manufacturing excellence. I knew from reading articles about the 30-metre boom and the keel that extended to 10 metres below the water to balance the 3,000-square-metre sail area. I also wanted to see the 2-metre turnbuckle that I had spotted in the article. When I finally saw *Mirabella V* it looked like a white-hulled sailing yacht. At 77 metres long, it did not dominate the horizon as *Sailing Yacht A* would one day do so menacingly. I was somewhat disappointed and busied myself with the preparation for the anchorage.

Then it happened: a 48-metre Perini Navi sailing yacht picked up anchor and turned for departure from the anchorage. Her position was such that as she turned, her heading was parallel and directly between my view of *Mirabella V*. Like one of Saturn's moons passing across the face of the planet, the significantly sized Perini Navi passed in front of *Mirabella V* and it dawned on me: the beauty of the design was the proportionality that belied the size of the yacht. This was a huge yacht, and the greatest achievement was that it did not look it.

Maltese Falcon joins this list as much for the vision of the yacht owner as the yacht itself. Built on speculation in 1989–90 rather than finding a buyer, the 88-metre Perini Navi hull was ageing. The risk lies with the shipyard when building on speculation and in this case, it had not paid dividends. It would have been a fair expectation that the hull may have wallowed to the point of scrap if it were not for Tom Perkins' interest in pursuing a rotating DynaRig system, his passion for designing the controls himself, and the ability of Dykstra Naval Architects to see it through. As it was, the yacht launched to great acclaim in 2007 and has received praise ever since. As with others on this list, it not only showed greatness itself but spurred others on to new levels of innovation.

It is an easy path to start with *Maltese Falcon*, add new levels of innovation and in a relatively straight line arrive at the 107-metre yacht *Black Pearl*. *Black Pearl* was able to capitalise on a global cultural and technological inflection. The world as a community wanted to see projects across all spheres that were built with an environmental sensitivity, and advances in power generation and storage facilitated this goal. To deconstruct all the innovations of *Black Pearl* would require an entire book, but in short the outcome was a 100-plus-metre yacht that can maintain all domestic services and cross the Atlantic using no power generation beyond the wind. Of course, 150 years ago this was common, but so was scurvy, and *Black Pearl* will drive others to follow and serve as the catalyst for more great design innovation.

SHIP CONSTRUCTION AND YARDS – THE NORTHERN EUROPEAN STANDARD

With such talent in design emerging in the 1980s, demand was placed on shipyards to raise construction standards. This was met and exceeded as shipyards across Northern and Western Europe took traditional shipyard construction quality from a process to an art. To visit a modern yachting shipyard today is to enter a creative environment where quality is the measure of performance. I have had the pleasure while representing clients to spend time in many of the leading yachting marques, and every time I leave with a greater respect for the production process that knits the pieces of a designer and naval architect's vision to the reality of construction. I am not alone in being impressed and fascinated by the process. I am often asked, 'Where are these wonderful creations built?' 'By whom?' And of course, the perennial favourite, 'Who is the best builder?'

There is no clear-cut answer to these questions. There are statistics and tables by country and shipyard, but these do not tell the story of the culture. Superyachts are more than the sum of their parts and the passion that starts on the designer's table and becomes a reality at the craftsperson's workspace is a journey, an artistic symbiotic relationship. Likewise, shipyards are also more than the sum of their parts: from a distance they look like one brand but pulling the veil aside reveals that they represent communities, both in and out of their physical constraints.

In addition to experiencing multiple refit periods I have undertaken two significant new constructions: the first in Plymouth, England and the second in Schacht-Audorf, Northern Germany. Although it might seem that each of these shipyards and constructions represented just one yacht in one country, in fact they provided a focus and a workplace for over 500 people every day for more than four years, the specialists being drawn as needed with no regard for geographical distance. Often, in both examples, as I walked towards the large shed where the hull was being transformed into a yacht I'd marvel at the diversity of businesses and nations that were brought together. The carpark was enough to make the United Nations blush in humility. A numbers-only table would also not show the impact to the community as the shipyards draw labour talent into their web.

Speaking of labour talent, I have worked alongside artistry in areas as diverse as steelworks, teak decking, lighting, electrical cabling, marquetry, loudspeaker placement and plant arrangement. These artists are independent contractors and are not listed when the shipyard table is published. They nevertheless represent the essence of the art of building a superyacht, the shipyard being the gallery that brings them together. This is not to diminish the role of shipyard; without it, these independent crafts would not exist. It is also to acknowledge that artistry spans many arenas; to the informed eye, a perfectly arranged electrical cable loom is as beautiful as a French impressionist painting, a perfectly laid teak deck is as jaw-dropping as a Joan Miró sculpture, and don't get me started on lighting temperature and colour. It is a diverse team that completes a superyacht.

Given this array of skills and people, you might wonder how superyachts can be built to and assessed by a common standard. This is where the 'Northern European standard' comes in. This is a contractual term used in construction agreements. It is difficult to define against a specific item, but in the whole, it represents an approach that over time has evolved to represent quality. It speaks to the great builders of Germany and the Netherlands. It represents generations of shipyard owners and craftspeople, and a pride that goes into every component and finish. To the industry it stands for more than a known standard: it represents a feeling when you are on board and a sense of security that everything will function as well on the first day as it will many years later. It is so ingrained into the seafaring yachting community that those fortunate enough to choose will often list working on North European yachts as a preference. To the owners, these yachts are beautiful and a reflection of their own brand; to the crew, they are reliable and safe when a long way from home and a long way from land.

Within this category are the esteemed builders of Northern Germany: Abeking & Rasmussen, Lürssen Werft, Blohm+Voss, Lloyd Werft and Nobiskrug. This is not an exhaustive list, and while these incredible firms differ, they all have a common thread: they build across commercial, military and yacht sectors, with each one providing development for the other.

Moving south-west from the Elbe, the Netherlands joins the category of virtuous and venerable yacht builders. With a history extending from the 1800s and bearing the Dutch Royal seal, the

Feadship Royal Dutch Shipyards that comprises both the De Vries and Van Lent families is a first stop, though these are not the only great shipyards in the Atlantic coastal nation. They are joined by Amels, Oceanco, Heesen, Royal Hakvoort, Vitters, Damen and another holding the Royal Standard, Huisman.

I had always heard the stories, but my time in the Netherlands came late in my career. I knew of the fabled yachts and from time spent living and working on a Feadship, I was respectful of the finished product, but I didn't understand the beating heart of Dutch shipbuilding. Then, in 2015 I was representing a client in a warranty claim at the Damen shipyard. Contrary to the mood of many warranty environments, the builder's team were gracious and even collegial in their disclosure. There was an agreed grievance, and the settlement proposed was fair to both parties. Having had experience in more aggravated settings I was somewhat wrong-footed by this. I shared this sentiment with my host and his response affirmed all I had heard about Dutch shipyards: 'Brendan, our name is that of a family and it is worth more than any one warranty settlement.'

We agreed on a plan for rectification and adjourned to the shipyard canteen for lunch. Having eaten at several shipyard canteens in the past, I was less than eager. However, this canteen resembled a high street cafe and the quality would have held its own anywhere. In the Netherlands, everyone is tall. At 180cm I am usually neither tall nor short, but in the canteen in Gorinchem I was staring at the shoulders of others. This is relevant since there was a commotion brewing in the canteen and all the diners were asked to stand to allow more staff to crowd the area. My host, Martijn, shepherded me forward to where I could see what was about to happen, and leaned down to translate what the speaker who had taken the microphone was saying. I don't speak Dutch and until Martijn assisted it was a confusing situation with many tall Dutch people laughing and jostling as though at a football match. A cheer went up as a silver-haired fellow holding a cigar came to the impromptu stage that had been set up to one side of the canteen. I didn't recognise the shipyard owner, but I immediately saw the influence he had on the workforce there.

This wasn't due to the fact that his family name was on every building and every pay cheque, it was more that there was a true fondness for him among the workers, in addition to deep respect for this shipbuilding patriarch.

The speech continued, as did the laughter, the words flowing quicker than Martijn could translate for me. A worker stepped on to the stage to great applause and was handed a wooden spike with his name on it. This was all very confusing, and it was not until the short ceremony ended that Martijn was able to explain it all to me. There are no privileged car spaces in this shipbuilding facility where 1,500 staff attend each day. It is part of the family's ethos that everyone is in it together and all are equally important to the goal. The wooden post I saw being given to the worker, alongside a private, more substantial acknowledgement, was in recognition of 40 years' service in the company. It entitled the holder to place the post wherever they chose and for this to become their designated parking spot.

I thanked Martijn for the explanation and complimented the company on an innovative recognition of loyalty and hard work. It was not until we left the canteen for the considerable walk back to the management building that I noticed just how many similar posts were visible. It seems a 40-year anniversary in the employ of this great yacht builder is not that uncommon. This little anecdote serves to illustrate the importance of the phrase 'completed to Northern European standard' within build contracts: it means a lot more than just an address.

Critics of my praise for the Northern European shipyards will speak to the volume and achievements of the Southern Mediterranean yacht builders in Italy, Greece and Turkey. They will also point to the ability in Asian shipyards to compete at the highest level and say that the USA has completed significant constructions over the years. I do not dispute any of these points and as my own experience grows, so might my opinion. Until that happens, I will draw upon what I am confident in representing and my 'great yacht builders' list will remain those who adhere to Northern European standard.

Key takeaways

- History and legacy matter. Make the effort to learn about the stories, the founders, the disrupters, the booms and the busts of your industry. You may think you come into your workplace without baggage, but everything that you see, do and feel is part of a legacy.
- Have fun looking in magazines, in books and online to discover your own preferences in terms of yacht design, then try to pinpoint what it is about those ships that stirs you. As well as being interesting and helping you to form your own opinions, it may also be useful one day in a job interview!
- Shipyards are fascinating places where diverse craftspeople come together to contribute their skills and share in the endeavour of making a yacht. Take some time to research how their process works. It will expand your knowledge of the industry and may influence which jobs you apply for. At the very least, you will be better-informed and keen to learn, which is always a good thing.

3

MEET THE PARENTS (OUR YACHT OWNERS)

To talk seriously about a yachting life, some sense of the yacht owners is needed. This is not a chicken-and-egg situation: there are no yacht crew without yacht owners. Before you ever had the idea to begin your best yachting life, someone else had to have a twinkle in their eye to commission the yacht to be designed and built and to take it into operation. It is therefore important to get to know these ringmasters of the yachting exhibition and in this chapter we will have a look at some behaviours – not specifics, but well-observed and researched generalities – of this rare breed of humanity.

This book does not dodge the fact that crew life brings with it many workplace pressures, and to understand where these come from is to learn about the guests that we crew are there to serve. It is difficult to convey to families and friends who are not in the superyacht arena the challenges crew face: the guests' demands are relentless and the days are long and mostly thankless. In addition to that, crew are separated from support networks and they often have very little idea of what is going on above their obligation to clean, maintain or operate. It is the owners who have invested their time, passion and funds into the adventure who have the ultimate oversight, so it pays to learn how best to interact with them.

I say owners are a 'rare breed of humanity', and indeed the stories that surround the guests, often reinforced by other crew and shore-based advisors, commonly result in superyacht owners and their guests being viewed with a distant alien awe. In my early days of yachting, I fell into this state too: I was so daunted by the owner that I stiffened and hardly made eye contact. I was projecting my earlier experience with senior military officers and did not allow the possibility of either I or the owner being able to relax. This response flies in the face of 'fun on the water' yet is exceptionally common. However, sometimes the 'accepted wisdom' of yachting needs to be challenged, and it is necessary to take the owners off their pedestals and learn to interact with them in a more open and comfortable manner while still communicating in an effective and courteous way that works for all parties.

PRIVACY – IT'S A CULTURE, NOT A POLICY

In a parallel to the reductionist term 'The Industry', those at the top of the grand undertaking of yachting are referred demurely only as 'the owners'. There is a dystopian perspective: is their influence so great that all crew, designers, builders and the myriad of support companies in yachting are in some way therefore referring to themselves as being 'owned'? The answer is maybe. However, the use of the term 'owner' also serves a different purpose: it maintains their anonymity and privacy. As you will discover if you work in the industry, yachting professionals speak among themselves in tangential terms about yacht ownership. There is little interest in the specifics of the actual owners – the dance is to complete the conversation without ever saying the name of the 'ultimate beneficial owner' (UBO) out loud. (The term UBO is more accurate than 'the owner', the latter being most likely a brass-plaque company domiciled in a remote jurisdiction rather than a person or family, though on board, 'the owner' is shorthand for the UBO, so that is what I'll use.)

A likely conversation between professionals might go as follows:

'Good owner?'

'Yes, nice family. They have been yachting for two generations and just use the boat for the family. Crew are looked after and there is a good budget to operate.'

And that is enough information. Professionally, there may be some follow-up questions about the owner's approach to crew and/or service providers, to ensure that the owning family are a good counterparty, but fundamentally, professionals do not care about the gossipy details of the rich and/or famous who own yachts.

It is a trip wire for me when a crew member or shore-based yachting professional starts telling personal stories about the owners. It is as though they have never read the unwritten rules of the game of yachting. I genuinely feel uncomfortable for them and highly recommend that you do not do it. Aligned with this is the equally terminal *faux pas*, 'I haven't shared this with anyone else, but I'll share with you now.' When I hear this, my mind translates this to, 'I will be equally indiscreet about anything I learn about this yacht, about you and about the owners to someone else as soon as I have the opportunity.' Discretion is a key skill for crew that will help you succeed in your chosen career; it simply isn't acceptable to gossip about your guests or what happens on board.

It is perhaps little wonder, then, that in the past ten years or so the depth and breadth of non-disclosure agreements (NDAs) required to be signed by crew has greatly increased. Personally, I don't have a problem with this in principle and willingly sign NDAs, since even when writing books about life in yachting I take pride in upholding someone else's privacy. The dark side of this adherence to privacy is that it has been used as an excuse to not report all manner of dastardly deeds on board yachts. Accidents, harassment, corrupt practices, pay disputes and more have in the past been hidden behind the veil of 'owner's privacy'. I do believe this to be changing, but if you have signed an NDA and experience something troubling, do remember that the NDA limits sharing information in a public forum, not in legal proceedings.

CANNERY ROW – A POTTED HISTORY OF YACHT OWNERSHIP

As with hem lines, physical yacht design follows fashion: the curves move to hard edges, the masculine intertwines with the feminine and the colour palettes move from pastels to greys to earth tones. In lockstep, the owners move through the parallel economic fashions. Yacht ownership can be used as a reliable metric of the movement of the tide of global economics and power. In the past, members of the royal courts used to sojourn to the coast and the ruling monarchs to their yachts – yachts that had been bought for them by the mercantile class seeking favour. This model has been amended over time to the situation we have today, where the ruling monarchs are often tech founders, though the protocol and the court that surround them continues with only slight amendments.

With the decline of monarchies and their dynastic rule in the 19th and 20th centuries, the ruse of buying yachts for the ruler was bypassed and the great traders instead bought their yachts directly for themselves. An example of this is *Sunbeam*, which has already been introduced. This yacht served as a representation of

Victorian-era English wealth accumulated by the Brassey family through their building of the railways that were redefining Britain in the 19th century.

In parallel across the Atlantic, the Vanderbilts were also making their fortune from the ever-expanding railways in the USA. Unfortunately, their 285-foot three-masted steam yacht *Alva* sank after a collision in fog in 1894. All on board were rescued and the misadventure did not dampen this family's attraction to yachting. The next *Alva* moved to diesel and at 16 knots was able to cruise at pace, complete with seaplane and gymnasium, when it set out in 1931 for a world cruise. In a twist of irony, having been built in Kiel, Germany, *Alva* was requisitioned by the American government to support the war effort and renamed USS *Plymouth*. It was then promptly sunk by a German U-boat in 1943.

A competitor of Vanderbilt was Jay Gould, a name (maybe/likely) borrowed by F. Scott Fitzgerald for the more famous but fictitious tycoon Jay Gatsby. Gould loved yachting so much that alongside building his yacht *Atalanta* he founded the American Yacht Club. The club was needed, since soon after the first nation builders began to place their wealth afloat the oil tycoons and the robber barons of the early 20th century discovered yachts. Names that remain omnipresent in culture and the industry today were all represented in yacht ownership: Rockerfeller, Carnegie, Duke…

Find a name, and find a yacht. Find a yacht, and find an incredible story. Some involve tragedy, especially when turned over to war service, and some follow zigzag journeys between owners. One example is *Nahlin*, which was built in 1930 and is still in fabulous service. It was built for a Scottish heiress and cruised with British King Edward VIII until, in 1937, when it was bought as the Royal Yacht of Romania. At the time it could never have been foreseen that only ten years later the monarchy would be overthrown by a communist regime and then that this regime would collapse in 1989 with a revolution to displace the brutal dictator Ceaușescu and *Nahlin* would be in a state of ruinous decay. Just when it seemed *Nahlin*'s last chapter had been written, the grand vessel was bought and restored to above its original by the visionary English engineer Sir James Dyson, who owns it today.

Seeing *Nahlin*'s story as an example, many of the pre-war yachts moved around between private and state ownership. Yachts served their time in war service, and many subsequently waited in a state of decay as the world found its economic engine again. This did not take long and yachting soon moved forwards like never before. 'Super', 'mega' and now 'giga' were prefixes that would be used to describe their escalating size and complexity and as the yachts changed, global events moved yacht ownership in the modern era. The USA leapt ahead like no nation ever had, the Gulf States began to swim in their oil wealth, the Soviet Union collapsed, the term 'oligarch' entered the lexicon, and the internet created wealth faster and at a never-before-seen scale.

With historical diversity lurching from monarch to industrialist, kleptocrat, dictator and tech founder, it might seem impossible to draw the median line that links all these types of owner. Yet some time ago I chanced upon the most fabulous quote in John Steinbeck's *Cannery Row* that seemed to bring together all I had observed in yacht owners and all I knew of others:

> It has always seemed strange to me [...] [t]he things we admire in men, kindness and generosity, openness, honesty, understanding and feeling are the concomitants of failure in our system. And those traits we detest, sharpness, greed, acquisitiveness, meanness, egotism and self-interest are the traits of success. And while men admire the quality of the first they love the produce of the second.

Now, referencing this quote could be read as me being the cynic and seeing the worst in those who have risen to superyacht owner wealth. This is not the case: I see the 'billionaire gene' often present in yacht owners as an incredible study. I often spot behaviours that are consistent with their success, but these behaviours are innate to them and are impossible to emulate.

> ## INSIGHT
>
> A question that is often asked is, 'What is the attraction of yachts to billionaires? Why do they all seek yacht ownership?' The answer is usually fairly simple: 'Because they can.'

THE BILLIONAIRE GENE

From the perspective of a reader hoping to succeed in the superyacht industry there is a need to understand the client. As I wrote earlier in this chapter, I recall being so unsure in my early days how to engage with the yacht's owner that I avoided eye contact and really any engagement whatsoever. In hindsight, my deference could well have been mistaken for disdain. Why I behaved like this I cannot recall now, but what I do remember is that existing crew had developed stories surrounding the owner that built a wall between them and newer crew. It turns out that this was a false wall, and its purpose was perhaps only to protect existing crew members' own places in the orbit. To counteract this, I now hear myself repeating to other crew, 'Challenge everything. Do not let rumour become a rule.'

There are consistent inconsistencies within the billionaire group, with the consistencies being greater than the variance due to the unique life that extreme wealth creates. Among several behaviours that are worth exploration, one that stands out is their need for information to be presented at its most reduced level – 'just the facts'. Reporting to billionaires in a manner that would work with a primary age or younger student is critical. This is often referred to as the Feynman principle after the great physicist Richard Feynman, who in essence said that you needed to understand a concept fully in order to explain it simply. (There's more to it than just this, so I urge you to look it up.)

'Just the facts' extends to supporting what you are saying with source data. Among yacht owners there is usually a tenacity for getting to the source (the first principles), and opening with phrases such as 'I think/I feel/Maybe we should...' without substantive supporting references tends to end terribly. In the introduction I mentioned the 'signal-to-noise ratio' in the context of radio propagation, and this is also my guiding thought when presenting to yacht owners. To retain their attention, you must deliver clear and valuable content with supporting references. If you meander or waffle, they will end the conversation and move on.

I have been well prepared for this type of communication from my time in the Navy, where presenting or speaking with the captain on a good day was a daunting prospect. The extensive preparation, inspired by anxiety of failure, ensured that junior officers cautiously made their approach to captain's briefings armed with highlighted notes to ensure all questions could be answered. After some appalling fails during these interactions, my preparation improved, but it never made up for the more than 20 years' experience the naval captain could draw upon; like the billionaires that I would go on to encounter, the captain's questions always exceeded my capacity to prepare. The yacht owner's ability to parse the background noise and come straight to the signal never ceases to astound me. I don't know how the skill is developed or taught – maybe it's just the case that their lives are filled with staff like me presenting to them, which has meant that they must develop a filter.

For someone making their journey and seeking superyacht success, then, the ability to present simply, concisely and factually is an essential skill to train. However, simply does not mean stiffly, nor does it mean speaking so bluntly that it crosses over into disrespect. In the yachting community there are laconic Australians, New Zealanders or Northern English who do not feel they need to change their behaviour for anyone at any time – and that can work if it is authentic and your natural mode of address. I have learned my own voice, and while I would quite like to emulate the cheekiness I see in others, I do not trade in the relaxed humour of my countrymen. I am more cautious with my words and try to develop the rapport through a truthful, almost flat presentation of information. How you convey the information will to some extent be personal to you, but in general, it is best to be courteous, cautious and professional.

KEY TAKEAWAYS

- Owners are essential to yachting. Without them, there are no yachts, so it makes sense to learn how to interact with them.
- When reporting to owners, make sure you have prepared well in advance. The information should be presented succinctly and be backed up with facts and supporting evidence. It is usually best to be authentic, for you and for the owner.
- Do not necessarily take everything you hear about an owner as fact. Sometimes other crew exaggerate traits or build walls. So long as you go in well prepared, with a 'just the facts' approach, and speak in a polite, concise manner, you'll likely be fine, and able to form your own opinion and approach to conversing with an owner.

PART TWO

LIFE ON BOARD

Here we go... Not all sections were created equally, and this is going to be the steak in the sandwich of this book. This is the bit where we place some honesty against all the myths of the social media-enabled versions of #yachtlife. Spoiler alert: not everything is going to be positive, but we are here to share truths.

4

First days on board

Having introduced in broad terms the history, the language and the owners in the superyacht circus it is time to approach the question, 'How do I get inside this tent?' As much as the yachting industry has grown by complexity, value and reach, the way to enter is still largely haphazard. My own path of 'a friend introduced me' remains valid for a large sector of the market, and like my own journey, many crew have no idea what they are really getting themselves into. In this chapter I hope to fill in some of the gaps and provide an insider's view on what it feels like to finally step beyond the Instagram version of yachting and embrace reality.

In my story, at the age of 30, I had packed my apartment in Sydney into storage and boarded a plane with a single bag. Unfathomably, with a logic lost to time, it contained rollerblades. Please don't judge me on this. With hindsight, I have this piece of advice (aside from not packing rollerblades to go and work on a yacht): avoid putting your personal goods in storage. Sell them, give them to charity, do what you must, but do not do what I did and end up paying five years of storage fees to then go through the dusty piles and realise that the once treasured items were mostly rubbish.

In addition to what you leave behind, what you bring is equally important: a crew member will be judged by their joining behaviour. This is possibly harsh, but like any professional environment there are many unwritten rules in yachting that provide stronger guidance than anything documented. Joining with only a collapsible bag in tow seems banal as a piece of advice, but nevertheless, the crew member joining with a large, hard-shelled bag will be mocked for their lack of awareness of the environment they are entering. But all this talk of what to take is getting ahead of the journey. Before packing those rollerblades, I needed to get the job in the first place.

INSIGHT

Begin yachting travelling light, in mind and goods; you are about to change your way of living.

THAT JOB ON A BOAT

Learning the backstory of yachting and some crew culture is important, as outlined in Part One of this book, but all that knowledge is of no value if it cannot be applied. This early in the book I hope there is a group of readers saying, 'How do I get me a bit of this?' My first answer would be, 'Don't follow my path.' I joined that first yacht because my childhood friend risked his reputation by putting my name forward for a position – more fool him, some would say.

Like so many, I entered the industry with little idea of what was in store and fell short of my own, my childhood friend and my colleagues' expectations. Among many who begin similarly misinformed, I say with confidence that I stood out as an outlier in poor performance. Many of those involved in my first yachting exposure more than 20 years ago have remained my friends and have all enjoyed telling me the impressions I made in those first days, weeks and, ahem, months as I fumbled through my infant days on board. It was not specific lack of maritime skills – I had grown up in boats, studied with the Navy and had commercial maritime experience. Where I was found wanting was in a lack of cultural awareness of how the crew interacted with each other. I was a deckhand, yet all my career references were derived from me being an officer in a job where such ranks meant something.

I did not have to do yacht-specific training because, as I said, I was already established in the commercial maritime industry thanks to my time in the Navy and commercial vessels. However, for crew joining yachting 'from the street' there is a requirement to complete an introductory course. The Standards of Training, Certification, and Watchkeeping (STCW) Basic Safety Training course is the required certification for anyone working on commercial vessels, including yacht crew. The STCW is a convention established by the International Maritime Organization (IMO). In a world where power sockets differ, it is delightfully universal in its content and its acceptance. The components are not onerous and unless done in the peak of winter, are enjoyable.

COMPONENTS OF THE STCW BASIC SAFETY TRAINING COURSE

Personal Survival Techniques (PST)

- How to survive at sea
- Practical training in the use of life jackets, life rafts and survival equipment
- Survival strategies, distress signals and emergency procedures

Fire Prevention and Fire Fighting (FPFF)

- Focuses on understanding the causes of fires on board and how to prevent them
- Practical exercises in using fire extinguishers, fire hoses and breathing apparatus
- The organisation of fire drills and emergency response

Elementary First Aid (EFA)

- Basic knowledge of first aid techniques

- CPR, dealing with bleeding, shock, burns and injuries that could occur on board
- How to maintain vital signs and stabilise injured crew members until professional medical help can be obtained

Personal Safety and Social Responsibilities (PSSR)

- Safety awareness and accident prevention
- Importance of teamwork, communication and understanding social responsibilities on board
- Safe working practices and the importance of complying with regulations

Proficiency in Security Awareness (PSA)

- Security-related duties and how to contribute to the security of the vessel
- Recognising and reporting security threats, understanding maritime security levels, and knowledge of emergency procedures related to security

It is a great course and a chance to gain a sense of what is going to happen in a non-visible future. The course is usually structured across five days and is offered in various key yachting centres around the world. At the time of writing the lead providers include, but are not limited to:

- Maritime Professional Training (MPT) – Fort Lauderdale, USA
- Bluewater Yachting – Antibes, France
- Warsash Maritime School (Solent University) – Southampton, UK
- UKSA (United Kingdom Sailing Academy) – Isle of Wight, UK
- PYT (Professional Yachtmaster Training) – Durban, South Africa
- Crew Pacific – Cairns, Australia

Entering this new world comes at a cost and it will be tempting to look at the training providers via the lens of 'Who is the cheapest and closest to home?' These metrics may be valid, but if possible, look to which provider is likely to introduce you to the culture of yachting and the social skills that will be needed, as much as which way round a liferaft should be. Also important – though something you have no control over – is your classmates, as some of the introductions made during this very first training week will continue throughout a yachting career.

In parallel to training is learning how to present yourself to yachts, your would-be workplaces. This can be daunting. There are two paths, and most new crew must undertake them in parallel. 'Dockwalking' is exactly what the name suggests – you walk around a dock that contains superyachts trying to get a job. This is tough, but a necessary part of the industry. New crew walk with their CVs in hand presenting themselves for either daily work on board (as dayworkers) or, ideally, that first full-time position. It takes a lot of courage to walk up to a large yacht with a cabal of crew who are all dressed alike on board and say, 'Hi, I am looking for work, could you help me?' I make a point of coming down and talking to dayworkers; rarely is there a role for them at the right time and place, but I want to respect their effort and gumption for fronting up and asking.

The second and more robust path is to register with a crew placement firm. Good firms mentor crew from the very beginning and develop them as a product that is appealing to a future employer on board a yacht (for more on mentors, see p. 200). As mentioned, I had my first role handed to me through my friend, but this came at a cost and I may have had a better start if I had researched more, learned more and spoken to more people before stepping on board that first day. If I had worked with the support of a professional recruiter, a 'crew agent', then I would likely have had a much better sense of how everything fitted together (and wouldn't have packed my rollerblades).

Crew agencies are the firms that take raw potential crew and try to shape their profiles and CVs into something a yacht somewhere, anywhere, can make use of. Like many service providers where there is not a high barrier to entry, their quality varies, so you do need to do your own research before signing up to one. It is best to work with a reputable firm, but when you are new, what this is is not easily defined. Some markers to look for include: years of trading, the firm having physical premises and MLC certification (MLC stands for Maritime Labour Convention). These metrics signpost a real firm, but still may oversimplify what is a nuanced business.

Get it right, though, and you'll be off to a great start. Good firms are highly respected members of the yachting community, and their senior staff are sought out for their insight. They should be focused on answering these common questions:

- Where do I start?
- What are the most common mistakes made by new entrants and experienced crew?
- Can I make a long-term career in yachting?

Some of the answers to these and other important questions can be found in the social media postings of the leading placement agencies, and are a sign that this is probably a reputable firm. Avoid the advertisements showing yacht crew having fun. That does occur on superyachts, of course – and if it didn't, why would you want to do the job? – but what you are looking for is a firm that shares the landscape truthfully. As with the real circus, it looks like a party from the outside, but inside the tent, yachting is often hard, tiresome work.

To hear from one of the firms directly I contacted Liam Dobbin of Wilsonhalligan. Liam and I have known each other for a long time, and I have been aware of the firm since its inception. The founder Terry Wilson showed faith and care for my career in a way that was and remains an outlier in the industry. Among some other very reputable firms, Liam and Wilsonhalligan are leaders in their field, so it is a pleasure to have Liam's words join mine.

Guest anecdote: Liam Dobbin – Wilsonhalligan

Recruitment companies and recruiters often have the reputation of being like a second-hand car salesman where the commission is the most important thing.

At Wilsonhalligan we look to dispel that reputation and even call ourselves recruitment consultants – using the definition of consultant as 'a person who provides expert advice professionally'. A person's career and livelihood is not something to be meddled with or gambled upon for commission. We look at long-term gains over short-term ones. The company's reputation for professionalism is on the line with each interaction with a crew member and person on board a yacht.

Wilsonhalligan started in 2006 from a maritime human resources background. Yachting was taking off at this time, especially with yachts 100m+. The importance of knowing the right type of person who suited yachting and onboard life was needed. This is where yacht and sea-going recruitment differs so vastly from mainstream recruitment: you are putting together a high-performance team in a stressful environment that is 24/7. The crew on yachts live and work together for months at a time.

Any person who registers or connects with Wilsonhalligan is a potential client; a green deckhand has the potential to be a captain in years to come. The business has a core value of being personable. Over the years we have become long-term friends and confidants with the crew we have interacted with. It has been a true pleasure over the years as an owner of Wilsonhalligan to be involved in building teams on board yachts.

A yacht and its crew have a personality, which is a combination of the owner's needs and the leadership of the captain to developing a culture on board. Our role in this elite maritime sector matches the crew to the yacht's personality. Knowing a crew member is going to fit into a high-performing 24/7 environment is a skill.

Leadership within yachting is growing at an exponential rate. Captains are akin to business CEOs more than master sailors. As superyacht captains deal with growing demands for operational efficiency, sustainability and crew well-being, leadership in yachting is also changing. To create unified and driven teams, modern captains are cultivating leadership abilities that combine technical proficiency with emotional intelligence. Leadership dynamics are changing because of an emphasis on diversity, environmental responsibility and ongoing professional growth. In addition to providing outstanding guest experiences, captains are also required to emphasise safety, comply with legal regulations and retain crew. Today's superyacht leaders will be able to manage the ship and its multicultural crew with professionalism and creativity thanks to this change, which promotes proactive problem-solving, flexibility and effective communication.

Training is at the forefront of the future in yachting: how can crew now be better equipped for changes for the coming generation of crew and owners?

In recent years multiple leadership trainers have come to the market, giving captains and crew the necessary skills to manage high performers. Crews' mental and physical health is now at the forefront of discussions and development for the future. Wilsonhalligan was an early adopter of crew well-being and mental health based on our own onboard experiences. One value at Wilsonhalligan is to encourage the current and the new generation of captains to move with the times and develop as brilliant leaders and managers.

Great onboard leadership in turn results in a more successful yacht, and crew retention is a byproduct for the yacht's owner and guests.

Personal anecdote: One day in September

Everyone who was at an age to be sentient on 11 September 2001 can remember where they were when the attack happened. They remember standing before a screen watching something that shifted their sense of 'normal'. Gravity is consistent, prices rise, casinos win and commercial aircraft do not fly into the tallest buildings in the world – until they did. My experience of this most historic day was on board a 70-metre motor yacht, berthed in Monaco watching a small screen that was used to monitor the satellite feed for the owner's television. It was my first day with this yacht and my first day in the world of superyachts. It was a strange day, to say the least. After my shift ended, instead of putting my best foot forward, I went with everyone else in the marina to the Monaco landmark Stars and Bars and drank beer like the world might end.

Monaco usually sits aside from the world's traumas. If there is a tragedy, Monaco will hold a gala to raise money; they approach life's crises in black tie and ballgowns. But 9/11 was different. On that terrible day it felt that tragedy could reach out and touch anyone. The bars filled and the billionaire sat next to the sports star, who sat next to the deckhand. In Stars and Bars, the deckhand was me and the sports star was a recently retired Australian motorcycle champion, foot still in a cast. We watched the talking heads on the television, the endless loop of the planes, and fell into an easy shout buying each other beer and wondering whether these planes meant the world as we had come to accept it was changed forever. In some ways it did and in many ways, in yachting particularly, it just kept on turning.

Since then, yachting has taken me on a physical and emotional odyssey that I could not have foreseen through the haze of beer on that very first day. I have circumnavigated multiple times and visited and become intimate with destinations I was not even aware existed when I discarded my comfortable Australian routines. I have worked within a global community and met world leaders and many of the world's wealthiest families and individuals. Even with all this, my experience is limited to a tiny slice of all that yachting has to offer. It is a crazy, diverse industry – and I wouldn't have it any other way.

LEARNING THE JOB

Having now agreed that yachting is indeed a special 'industry' and that all that are sailing today are upholding hundreds of years of legacy, it's time to take your first steps on board. For some, this may be your first ever job; for others, it might be your

20th. Either way, you are a newbie on board and the same rules apply. To best illustrate what to expect, I'm going share the story of my first days on board. Others will have different tales, and be starting from different places in terms of their prior experience, but hopefully my words will be instructive as an illustration of what to do, and what not to do.

Prior to embarking on my yachting career, I had, by my own humble measure, achieved a level of distinction in personal and professional endeavours. I was and remain proud of my ten years serving with the Royal Australian Navy. Then, having left the Navy at the age of 27, I ran the Northern Australian operations for P&O Maritime – which is also something I look back at with a sense of accomplishment. This was my history and my baggage, but once I was on board that yacht, honestly, nobody cared. I had been employed to wash a boat and that was that. I was told bluntly by the chief officer that deck crew were not welcome inside the yacht during working hours; like dogs on a farm, they were to stay outside and work until mealtimes, when they were briefly allowed inside. I resented this and felt a strong sense of 'Do you know who I am and all I have done?' But he did know: I was the deckhand, and he was my boss. He and I remain good friends and I see him now as having been a professional in yachting when this was a very rare commodity.

It was maybe my third day on board and the yacht was preparing for the small relocation from Monaco to Antibes. As dictated by my prior experience, I went to the bridge and was ready to assist. In an awkward moment the captain came on to the bridge and just looked at me, not understanding why I was there. The bosun was passing through at the same time and, noticing the awkwardness, in an act of kindness tapped my elbow and said, 'I forgot to tell you, you're on the aft deck for the move.' I left the bridge and walked aft, not sure what I was meant to be doing back there either. The yacht moved without incident and I quickly learned the deck routines and the handling of lines to meet the needs of a large yacht. They were not that different to those on board the tugs and barges I had recently left.

What did come as a surprise was the care and maintenance of paint and varnish. It seemed an obsession in those early days, and involved a language I had not yet learned: the products all had shorthand names, and the artistry needed for what seemed a tiny repair was of a different order of magnitude than I possessed. So I listened, learned, watched and followed, and after a while, I felt like I was improving. The chief officer must have sensed I was getting the hang of it as he then detailed me to wash the hull on my own. The exhaust smoke had put a grey sheen on the otherwise brilliant white hull, and it needed to be washed off as soon as possible so as not to embed itself into the paint permanently. I set off on a small raft kept for the purpose and had assembled everything I needed: a soft brush, a pole and some soapy water. I brushed and brushed but did not seem to be making any headway. I paddled myself to the ladder at the back of the yacht, climbed on board and went to the deck cleaning locker. I needed to increase my firepower, so I shuffled through what was on offer and found a brown scrubbing pad that was a 3M product with the formal product name of Doodlebug. It looked like just what I needed, and I had a bounce in my step as I returned to my raft. I was going to show my deck team colleagues, the bosun and the chief officer that I had crossed the commercial Rubicon and was ready to be counted as a 'superyacht deckie'. Armed with my new Doodlebug I once again attacked the exhaust residue on the hull, this time with fabulous results: the grey was coming off swipe by swipe.

A polyurethane paint system is built up of layers. Starting from the inside, it begins with a barrier coat to raw steel, filler to smooth the profile, base coats to build thickness and finally a gloss finish that forms a hard 'skin'. Done well, it delivers a flawless high-gloss finish, making a yacht look incredible. The challenge is that it is expensive to apply and difficult to maintain and repair. Later in my career I would be responsible for a polyurethane paint project quoted at €16 million for the hull and house, and I became quite the expert. Unfortunately, on this day I did not have this future knowledge to draw upon on my washdown raft. Due to my excessive scrubbing, I not only removed the exhaust stain, but also the gloss for a section 3 metres long and 2 metres high. A repair of this size would go into the thousands and never be quite right. Within the deck team was one archetypal old-school Australian yachtie, 'Jonesy'. Jonesy was a skilled shipwright and could do anything on deck with ease. He had already saved me many times and now he leaned over the side to look down and see how I was getting on. He saw me smiling back, scrubbing the hull with the Doodlebug. In his quintessentially laconic Australian drawl, he said, 'Mate, just walk away. Seriously, get off the raft and get the fuck away.'

As good as this advice was, I had to tell the bosun what I had done. To his credit, he took full ownership of the error for not having given me enough guidance. He was then, and remains to this day, a fabulous fellow who always sees the best in people, though he is something of a rarity in yachting. I am disappointed to say that there can be a tendency to not make room for the education and development of new entrants, to see inexperience as a weakness to be exposed rather than a space that needs filling with knowledge. As it turned out, the hull was to be painted in the coming weeks and instead of making a great show of my disaster, his response allowed it to just become an anecdote that stayed with me in the crew: I was 'the Doodlebug guy'.

My takeaway advice from this story, in addition to not using a Doodlebug on a polyurethane paint system, is that if in doubt, ask before you act. If you're unlucky, you might be met with impatience, but that's a reflection on them, not you. It's much better to annoy someone but get things right than to plough ahead blindly, as well-intentioned but poorly informed crew can cause significant damage.

LEARNING TO BE SOCIAL – THE BLUE LADY

Professional competence in yachting remains closely followed, or at times, led by social loquaciousness and being 'fun in the crew'. It was made clear to me that my disastrous professional *faux pas* could only be settled by me buying the deck team drinks. This was formalised with a 'deck fine' system that I seemed to activate often and everyone else was immune to. Following my cleaning catastrophe there was talk towards the end of the working day across the crew radios of 'Brendan's shout at the Blue Lady'. It was fair to assume the Blue Lady was a bar, but beyond that I was completely unaware of the plan. My own idea was that I was going to go for a run and explore the Cap d'Antibes headland after work.

When I joined the yachting community, and wanted to 'get on' and make a career out of this new sector in maritime. I had partied in my Navy days and did not see the need to keep doing it. This made adapting to crew life a little harder than it needed to be: I was seen as stiff and not really getting into the yachtie way. If I could play the record again, I probably would have loosened up a little, but I saw (and still see) yachting employment as a privilege and something not to be squandered. If I was too stiff, then over the decades since I have seen too many who have been too loose. It may sound like a pithy public health announcement, but for a new entrant to #yachtielife, defining what your responsible relationship to alcohol will be is a critical choice. Being the last one home and the one with the most colourful night-out stories will make for fun morning tea conversation, but fundamentally, it is a short-term play. My experience is also of another time. I work alongside a team now whose social playground is not the bar, rather the before-work running group. They live the 'early to bed and early to rise' life and are very conscious of diet, exercise and their personal growth. I love to see this, but I do miss Jonesy.

LEARNING TO EMBRACE BEING NEW – GREEN IS GOOD

The term 'greenie' for a new entrant to the superyacht crew community is relatively new. In my first days, I and others were all referred to as FNGs ('Fucking New Guys'). I learned this when

I was shown the list of jobs in the deck workshop, which had the heading 'Shit Jobs saved for the FNG'. True to its title, the jobs reserved for FNGs were less than appealing, though it became a point of pride to me to complete them. My most recent maritime work had been on tugs and barges and in comparison superyacht 'shit jobs' were not that bad. I have to say, though, that the sense of entitlement and superiority from those only marginally ahead of me in yachting and less experienced in maritime industries did disappoint me. I knew I needed support to cross the threshold into yachting, but it did not need to come with resentment.

Sadly not all carried the grace or the sense of ANZAC camaraderie 'Jonesy' extended to me. Fortunately, FNG as a term has been lost to time, but the negativity that is projected towards new starters often remains. On one level this can be explained by the increased burden placed on other crew by having to carry a suboptimal performer, or maybe it is a defensive position, as a lot of new joiners bring a positive disruptive energy into a team and challenge the accepted, though potentially inefficient, practices.

Throughout my FNG/greenie days I continued to put extra effort into my work and my learning. The term 'extra effort' needs to read in the context of the time, when treating yachting as a professional career was an outlier choice. For most people, crewing on a yacht was good fun and offered a chance to travel and get paid. Performance expectations by crew and guests alike were a lot less than they are today, and these factors combined to create the environment where opportunities fell my way quite easily early in my career. This would not be the case now.

Today, it is a competitive market with most crew having a level of interest in their careers that far exceeds my own efforts of 20-odd years ago. Modern crew look to their leave periods as opportunities for further training and benchmark their progression against those they recognise as peers. Most also know exactly what they seek to achieve through a yachting career – again, far exceeding the knowledge I and my peers had a couple of decades ago. So, to succeed today, you need to do your research and really think about what career you want in yachting: what you want to achieve, where you want to go, and why. When you're doing the shit jobs as a greenie, having a sense of purpose will help you maintain your morale.

Purpose and fulfillment can also lie in doing a good job. I was used to hard work, so I found I was revelling in being a worker bee on the yacht where each day I was told what to do and did it. It was menial work, but there is a deep satisfaction to be found in simply washing a boat and seeing the beauty of the shining paint, the pristine teak and the clean windows. This professional enjoyment and sense of pride took a hit, however, when, quite early in our relationship, my now wife and I went to Austria to visit her family. My German language was not great, but I managed to pick up the words 'Brendan ... boat washer' as Yvonne was talking with her parents. Later when Yvonne and I were alone, I had to ask, 'When you were talking to your parents, what did you say that I did?' Yvonne replied, 'I told them you washed the boat all day.' Yvonne was correct: the fundamental job of a deckhand is to wash the yacht, set it up, pack it down, and repeat. I was just not ready to be deconstructed so accurately and at such a base level. I carried the ego of earlier achievements, but only I could see the baggage; to everyone else I was indeed a 'boat washer'.

Since this time, on multiple yachts and across all departments, I have seen countless examples of crew who join yachting bringing their sense of self-achievement with them and then fall heavily when the reality hits home. The deepest truth is that work on a superyacht is mundane and repetitive; success in the environment comes from making peace with this. When you are next in your capital city of choosing, maybe to gain your first visa to join a yacht, pause in front of one of the ubiquitous expensive hotels and think, 'Do I want to clean the rooms, wash the windows, sweep the walkways?' 'Yes' has to be your answer if you want to work on a yacht. If providing a beautiful environment for others does not give you joy, then save your money on the training and the visa and head home, as this is the truth of yachting.

INSIGHT

Yes, it is true, most jobs are repetitive and mundane, so what makes yacht employment different? The difference is that most jobs do not purport to be something else. The glamour that surrounds yachts veils the hard truths of the job. If you can embrace that, then you can make a go of things in yachting. If not, walk away.

KEY TAKEAWAYS

- Research exactly what qualifications you will need before you can embark on a career in yachting. Unless you have valid prior experience, you will need to complete a STCW Basic Safety Training course. Look for one that talks about onboard culture as well as the compulsory components.
- There are two main ways to get a job: dockwalking and via an agency. Dockwalking is emotionally challenging, but it does work. If you solely rely on an agency, take the time to research various agencies to ensure they are genuine. They should have a physical address, years of experience and MLC certification. One you have chosen one, a crew agent should help you hone your CV, provide advice, answer questions, suggest training and help you secure an appropriate job.
- Leave your ego at the door when you start a new job. Nobody cares what you did in a previous life. You are there to perform a specific function, and you need to be prepared to follow orders.
- If you are unsure how to perform a task, it is much better to ask than to guess, even if you are met with impatience.

- Getting the balance right between being 'one of the gang' and preserving your health and values can be difficult. Bar culture is certainly part of yachting life, but you should not be bullied into doing anything you do not want to. Being clear about your boundaries from the outset is a good strategy. If others don't like it, then that's their problem.
- Your time as a greenie is precious: embrace it.

5

MAKING A LIFE AFLOAT

Thankfully, superyacht employment provides opportunity (if you take it) for accelerated learning and experiences; you will not remain the greenie for too long. On board, you will often be replaced as the newbie within weeks, as a new person follows you into the crew. You'll move from nervously opening and closing cupboards looking for breakfast cereal to supporting others to find obscure items in even more obscure storage areas. With this knowledge of where and what gleaned, the next stage is the who and how of making a successful life for yourself aboard a yacht. To do that, we need to take a look at group dynamics. Superyachts place multiple humans, largely strangers, in a strange environment and take them to strange lands. This unnatural state of affairs inevitably causes stresses and strains, so what follows is my take on what those are and how to successfully navigate them.

LEARNING ABOUT CREW – 'OTHER BABOONS'

In order to do this, I'm going to talk about baboons. Bear with me, baboons are not our closest primate relative, but like humans, they dominate their environment (destructively) and live in social groups of up to 100 members. These social groups are complex, with relatives and non-relatives all vying for status and survival. This and many more parallels awoke the scientific community to the importance of their study in Darwin's time and ever since. It may have also helped the cause when the young Mr Darwin said in *Notebook M*: 'Origin of man now proved. Metaphysics must flourish. He who understands baboon would do more towards metaphysics than Locke.'

Among the scientists who have studied baboons are husband-and-wife team Dorothy L. Cheney and Robert M. Seyfarth, who in 1992 began a 16-year study of the social interactions of baboons in the Okavango Delta in Botswana, and subsequently published their findings in their academically rigorous book, *Baboon Metaphysics: The Evolution of a Social Mind*.

'Origin of man now proved. Metaphysic must flourish. He who understands baboon would do more towards metaphysics than Locke.'

There is intrigue throughout the book as the authors follow the baboons through lovers' quarrels, social climbing, greed, incest and even to murder, and thoroughly examine how much of the baboons' behaviours are instinctive and how much are driven by actual thought. I read the book, all 358 pages of it, and then read an interview with the authors. One phrase in particular struck me. For all the years and all the data analysis contained in this significant book, Seyfarth replied to the question, 'What are the biggest challenges to baboons?': 'Any way you look at it, most of the problems facing baboons can be expressed in two words: other baboons.'

It was a beautiful answer, so succinct and clear. I have used these two words – 'other baboons' – many times since to explain crew interactions. The challenge of the baboons remains valid to humans, even though we have (mostly) progressed, even though the onset of reality TV would have Charles Darwin questioning his theory of evolution. As a constructor species we developed physical tools to make fire, then the first spear and a little while later a 100-metre yacht. This is just one facet of our skills, though. As a social species we also need tools to overcome our own challenges of working in ever-increasing team sizes.

Taking those first steps in yachting requires STCW introductory training – as introduced in Chapter 4 – and in no way am I reducing those skills: firefighting, sea survival and first aid knowledge are critical and may save lives. But learning about the cultural and social environment on board will make every other day – every 'normal' day – that much better. This is why, when choosing your initial training venue, you should check whether they talk about social interactions and the stresses of onboard life; the better schools will do so, and it is for you to probe to find the answers you need. Most interviews through your career on yachts, at all levels, will be primarily focused on your skills and experiences: Have you worked in a restaurant, a hotel, a watersports centre or as an engineer in a related area? I am not discounting that these skills and experiences are critical for your ability to perform the functions you will be paid well for, but they will not be your only challenges in the role. As for the baboons Cheney and Seyfarth studied so diligently, your challenges will be 'other crew'.

Issues are not due to any single personality or cultural trope being present in the crew. In part, the problems are caused because on board a yacht the personal space people get in a shore-based life – when they go home – is absent. Behavioural idiosyncrasies – yours and others' – that normally play out privately, a long way from the workplace, will be part of everybody's everyday life. At its most basic level, the personal hygiene and eating habits of your fellow crew will enter your thoughts to a degree you would prefer they didn't. I usually put this question to potential crew early on in an interview: 'Where do you go in your mind and who do you have to call when you have a bad day on board?' This is not a hypothetical scenario – the response needs to be thought out in advance. Those who come into yachting as an island unto themselves and without thought to who and what is their support strategy often falter when their normal references are removed and replaced by the strange, confined and structured environment on board.

Keeping the theme of baboons foremost, the communal areas bring out the best and worst in crew: the kind, the conceited and the clumsy all find their space in the crew mess. Not to get too far ahead of my personal story, but when I moved along my journey to captain, my judgement of crew was formed as much by their behaviour in the crew mess as it was from demonstrated skill in the roles they were hired for. When I mention 'communal behaviour' to a crew group, I know everyone is looking around as though it is a problem caused by 'someone else'. It is inconceivable to them that their own behaviour could cause offence, and yet everyone has their flaws.

Communal living in a very confined environment where no allowance is made for personal space is testing. A normal routine of home/workday/home (or the now common home/work from home/home) allows for the 'space to spread'; on board, this luxury is removed. In this environment, personal items left in communal areas represent a yachting sin of the highest order. While even writing this seems somewhat banal, it isn't, and not everyone can adapt. There is a need for crew to always keep themselves 'together', to not allow their personal zone of influence to impact the lives of others. This is harder for some than others and not everyone can survive, let alone succeed, in the realm of yachting baboons without a) feeling compromised in their own needs or b) upsetting the others in the troop. So, take some time now to stop and really think about whether you could cope with this. If the answer is 'No', then life aboard a yacht will not be for you.

I am projecting my own experiences into every story, but the situations and events are common to many making their way into the yachting ecosphere. I joined my first superyacht as a fully formed adult. I had professional and personal experience and I was used to having agency over my daily choices. I would go so far as saying I had been the master of my own domain in my shore life, but this autonomy is not what was sought once I was on board. What the yacht wanted was for a function to be fulfilled: it wanted the boat washed, the tenders driven and the anchor lifted and lowered. It did not want great initiatives from me. The processes on board were not there to be reinvented – there was not the time, the will or the need.

My humanity, my personality, was also not really needed. As Yvonne so concisely pointed out, I was a 'boat washer'. If before I had been able to choose my own soundtrack, be the DJ for my own life, the beats were now laid down by the yacht and those established within it, and that was what I had to listen to. My story was washing boats, but it could have been cleaning rooms, serving meals or fixing engines. Whatever the role, the work needs completing – and yachts create a *lot* of work. Yes, there will be incredible travels, friendships and fun, but these only occur after the work is done. The crew who miss this priority will be forced out of the troop, and quickly.

Key takeaways

- There is minimal personal space on board, and communal living is 24/7. You need to be someone who can cope with this, both on a practical level – keeping your belongings tidy – and on a mental level.
- In order to keep yourself together, you need to know in advance who and what your support strategies are. Who are you going to call when the going gets rough? What coping strategies have you got in place to handle stress?
- Life on board involves marching to the beat of the yacht, not your personal soundtrack. Are you someone who can follow orders? Can you do a job without bringing your ego into it?

DISCOVER YOUR OWN COMPETENCE

Knowing how to maintain personal hygiene and keep my cabin clean was a good start, and my time in the Navy's version of communal living, stacked three high in a 70-person sleeping space, helped me adapt to life on board more easily than many. I still bristled internally at the loss of perceived control of my life, and I was still boat washing, but I had started late in life and had somewhere I needed to get to. I was heading to the bridge and needed to take some steps in that direction. As much as yachts have increased in size, diversity and capability, the manning tables for the crew operating them have remained relatively static. There are more crew and more specialists, but the fundamental path remains the same. My experience speaks to my journey on deck, but it transfers happily to engineering and the hotel teams, so is valid for people in any role.

I worked the allotted time in the appropriate roles to step forwards through the accepted metrics of yachting career progression: time and certification. The fact that I already held more certification than was required for the role gave me a head start, and while I needed every day in each position to gain experience, I was often asked by other crew, 'Why aren't you applying for jobs [at the next level] now, with your qualifications?' When I replied that I wasn't ready and needed more time, there was often a somewhat dismissive shrug.

There is a competitive nature to career progression around the yacht fleets, as peers compare to peers. That the person who sat next to you on that very first week of STCW training is now two levels ahead of you creates career envy and causes many to leap to the next role before they've really looked. There are often 'time-sensitive vacancies' available, and if a crew member is ready to step quickly, they can advance at an accelerated pace. This is not an invitation to do so. It is a caution *not* to. You are the person who needs to place a brake on your own career acceleration. Self-awareness is a tricky thing, so find some key workplace performance metrics to help you assess yourself and your readiness to step up a level. I cannot give these to you, but you need to objectively assess the people in the roles ahead of you and what they do daily, and ask yourself, 'If I were put in a new environment, could I learn the environment, complete the higher-level functions of the next role and lead others simultaneously?'

The reason you must make this assessment yourself is due to a quirk of yachting in that most surprisingly, core competence is rarely covered in interviews or references. This may be disputed by some, but I have given many hundreds of references, and questions about core competency are rarely asked. Once the certification and time in role have been verified, the inevitable next employment question is usually 'Were they a good team player?' (whatever that means). I often introduce role competence and observed leadership into the conversation, but it is generally acknowledged and then dismissed. I think I am viewed as a pedant who misses the bigger picture when I offer information that the 'head of service' applicant was not competent making a coffee or that the 'deck officer' struggled with basic collision regulations for shipping. It does speak to yachting's priorities that social acceptance is prioritised over the answer to the simple question 'Did they display core competence, and did they develop this in the period you knew them?' If this paragraph has you rubbing your hands with glee and your path to promotion just became flatter, then you are not ready. 'Ready, fire, aim' or 'Run fast, break shit' may have been the mantra for tech firms for some time, but they are building applications and are not responsible for the lives of others at sea. In contrast, you soon will be and being competent really does matter.

To avoid being an imposter in the jobs you covet you need to stay in one place and really learn the role. This is good news, then, as longevity is the yacht recruitment holy grail. In contrast to many other industries, the measure for acceptable 'longevity' on a yacht is measured in months, not years. Yachts simply have higher staff turnover than many shore-based industries. There are various reasons for this, which you will learn for yourself, but the takeaway is that there is a very easy way for you to stand out: stay in one place for a couple of years (minimum). In my own story, a lack of self-confidence to seek change kept my working on my first yacht longer than most, to the point where I moved from being the FNG to one of the longest-serving crew members. Over four years, I progressed and I became a second officer – a 'demi HoD'. I was not a full head of department, but nor was I a worker bee. The sphere of a second officer is a great kingdom to roam, a land where there is sovereignty over daily work, yet not to the level where responsibilities are a burden.

My extended period in this 'zone of peace' did give me increased competence, but where it helped most was in handing me the longevity chalice, which few others shared. This gave me an advantage. That said, even though I spent a good while as a demi HoD, I moved from greenie to experienced deckhand, bosun, second officer and eventually chief officer at a good pace. My holding back when others pushed me to progress was not enough and I still moved faster than I should have. My time through the deck ranks coincided with a time of high crew demand and in hindsight and with reflective honesty, I moved on from each role as competent, but I certainly did not begin each new role that way. I was living the 'Ready, Fire, Aim' approach that I have just cautioned against, and depending on how you face the world, you will either see this a cautionary tale or point of motivation. I leaned into each opportunity, backing myself that I could learn at a pace that would allow me to reach a threshold of minimum competence before my shortcomings would be observed. If you are an observer of others, even as a new entrant to yachting, it will not take too long for you to see that my approach is more the norm than an outlier.

KEY TAKEAWAYS

- If you can cope with life aboard and find you are thriving, it will not be long before you start to think about career progression. Remember, knowing your why and having purpose are essential aspects of dealing with the less exciting aspects of working on a yacht. Research what you need to do to make it to the level you hope to attain. What are the steps along the way? What qualifications and experience do you need?
- You are in charge of your own career progression. You will need the correct certification and time in role in order to apply for the next role, but the rest is up to you. You need to assess yourself and whether you are ready to step up. It pays to not rush on to the next stage before you are ready.

WHEN MORE IS NEEDED – THE 'CORNFLAKE CAPTAIN'

In only a few pages I have moved from longingly thumbing magazines to my first days on board and my time as a demi HoD. So far, everything had progressed normally, but what happened next was anything but. As a first mate on an 80-metre yacht I had an arrangement with my captain, largely unspoken, that I would do all the work and he would give me an opportunity, a leg-up, by supporting me if he was ever asked to recommend a captain. Like so many roles, the only way to gain a position as captain is to already have one – this is harder than it sounds in a competitive market. There are bars in the South of France, Spain and Italy full of all yachting ranks talking large of being ready for the next step but being overlooked, how they really run the show, and how the captain or their head of department does nothing.

In my story, the captain gave me the opportunity to walk the talk, by recommending me to step up from chief officer to being the captain on an Atlantic crossing. This was a big shift, so a 'babysitter' (recently retired) captain was hired to be on board as support (thanks, Rick). The ocean crossing finished without incident and with its completion I could list 'captain' on my CV. What happened next was laid out in my first book, but in short, just six years after not knowing how to clean paint I was hired as captain on one of the world's largest and most complex yachts. I still knit my brow at the boldness of the yacht's management and incumbent captain, now long-term friend, who saw something and took a risk hiring me.

My far from auspicious start of stripping paint from the hull was behind me and with more than a fair handful of luck I had moved swiftly through my yachting learning period. My exposure to more experienced crew that were free with their knowledge sharing was then and remains terribly rare. Knowledge is a tightly held commodity, sadly: too many ring-fence their careers by using opaque practices and restricting information sharing to protect their jobs. This practice is not unique to yachting – middling performers the world over have been honing this approach for as long as workplaces have existed. That this practice happens within an industry whose growth has defied the gravity of economic cycles, and where lives are literally dependent on knowledge being shared, is very concerning. We really should do better. I hope we do in the future.

With the support of mentors, my quarters had moved from the lower deck to the particularly large captain's cabin. As I unpacked into this cavernous new accommodation, I was the only person on board who knew that until recently I had been sharing a four-berth deckhands' cabin. Maybe it was hubris, but I didn't feel I needed to apologise for my situation, as with some level of confident humility I felt I deserved at least the opportunity to fail. Blessed with significant energy reserves, what I lacked in core knowledge I knew I could make up for with a level of effort, and through this, learn faster than anyone would notice ... or so I thought.

The 'Cornflake Captain' is not a title sought by anyone at sea. That I was being called this behind my back, by those I thought respected me and whose support I needed, was crushing. My engagement direction, my mandate, was to move the culture away from a crew too focused on what the yacht could deliver to the crew, to one that was aligned to the experience it should provide for the *guests*. It is easy to fall into the trap of being so consumed by managing the day-to-day that you forget the customer, and this situation is rife in any service environment. The physical yacht needs constant attention, and like a screaming child it is hard to ignore, causing the crew to feel, wrongly, that maintaining the administration is their focus and the guests are a distraction. Continuing the parallel to parenting, it takes a tuned ear to determine what needs attention and what is just whining.

By attempting to redress this I was covering ground and implementing a cultural change to reflect what had been asked of me at interview: to put the guests first. But the crew using the epithet 'Cornflake Captain' also spoke a truth. I ran around the field of play on board constantly chasing the ball and usually arriving among, or just after, a scrum of others. I did not have the experience to get in front of what was happening. I now speak of 'feeling the beat in the music of the yacht's day and moving in time with it' – but it takes time to sense this rhythm. your early stages in a role, when the beat of the yacht is not yet visible to you, each day it sounds like the orchestra members of the yacht are tuning their instruments all independently. The only solution is to do the hard thing and to say out loud, 'I am inexperienced and while I will work hard and will keep you safe, I need your help to perform at the level you and the yacht owners deserve.' This is written from my captain's perspective, but it scales to all positions.

It is comical to me now, with additional decades of experience behind me, that when I join a new yacht, I can tell crew that I need their help and I say it with ease. I still benefit from and require the support of others to reach my optimal performance quickly, even though in contrast to these early days I find I am uncannily able to be out 'where the ball is soon to be'. There is now a longer and more peaceful pause between event and reaction – I don't need to rush to make decisions because I am ahead of events. Later, I will describe some of the tools that I have found support this transition. Those that work for me may not work for everyone, but they point to the importance of having something to guide decision-making.

When asked about the motivations for my first book I have a rehearsed and reasonably honest response: 'I share my experiences so that others may benefit.' The reasonable part of the honest comment is that it is probably why a better person would write stories with an educational context for others in their professional sphere. My truth is not as noble: I write of personal experiences and learning outcomes to control the narrative and to hold myself accountable to the story. If it were not for the accountability that the writing demands, my default resting place might shift back to being the Cornflake Captain, of skating through by the skin of my teeth, just expecting nothing to go wrong – the inanely smiling member of life's audience applauding the opening night but missing the point of the show.

HOW TO WORK AND LIVE – YOU SPIN ME RIGHT ROUND, BABY

Alongside 'longevity', another key word in yacht employment is 'rotation'. My first exposure to rotation was linked to that first 'cornflake' captaincy. It is a word that, at least in the yachting vernacular, is as damaged as 'sustainable', 'eco' and 'hybrid' are in the wider world. Rotation is used and misused in a binary sense to describe crew employment patterns: roles are either described as 'full-time' or 'rotation', though neither actually reflects the work environment with any sense.

Due to the asset value of yachts now running into the hundreds of millions of euros, it is not financially responsible for owners to stop their operation just so that key crew members can take holiday. Overlaid upon this is also the concept of 'minimum safe manning', which defines how many crew must be on board and available to safely operate the yacht. And overlaid on that is the fact that by the time crew rise in seniority, there is a chance they will have started a family and therefore have commitments on shore that require their presence. All of these factors combine to create an environment where to achieve 365-day coverage there needs to be 'job sharing' or circling back to where we started, 'rotation'.

There was a time when yachts were smaller, and their owners did not notice that they spent a lot of time loitering in ports near where their crew lived. For their part, crew joined for the adventure and never expected to spend long periods away from the yacht. Relationships formed on board and the crew moved as one through their time; it is not hyperbole to say that the yacht was a surrogate home. I remember this era fondly in my own career and at times question why I left the smaller yacht sector, where routines are gentler and time is not monitored so heavily.

Given that I had commercial (unlimited tonnage) qualifications and entered the industry just as yachts were increasing in size, it made sense that I would drift in the direction of bigger yachts. I was not alone: other new entrants were drawn to the flame. Bigger yachts required more and higher-qualified crew. The logical first place to look was the cruise ship industry. These crew were important because they had the experience to bring the training and safety processes to a scale not seen before in yachting. What they also brought was a more transactional approach to their employment: where yacht crew historically were there for the journey, the commercial crew were there to earn income and holiday and go back to their homes. This mercantile approach caused some of the love to fall out of the room as owners realised their crews were no longer on board for the story, they were counting the days until they 'rotated' out for leave. This is now the norm.

The frequency and structure of rotational employment will vary, and I will not deconstruct all the types here. For new entrants it is enough to know that you need to assess the leave conditions for each role, and forget the routines of your previous life where weekends, public holidays and annual leave are fixed. In the wider community, holidays are based on a known calendar, onboard crew need to accept a shifting calendar to make the running of the yacht possible.

Getting the balance right between meeting the needs of the employer and your own needs is tricky, though. Like many of my yachting peers, I failed to put boundaries between what I could give and what the yacht (or its owner) could take. I opened the book speaking of just how wonderful the yachting industry is – and it can be – but as with many 'too good to be true' environments there is a dark side. The yachts and the owners draw you into their web and most (such as myself) fall, like Alice, willingly into this looking-glass alternate world. I speak with fondness of an earlier time in my career, a time when my life revolved entirely around the yacht's sailing programme and everything was in a carefree balance. I speak of the crew of that period and how close we all were. However, that way of working was not 'normal' in any sense, and it couldn't continue indefinitely. That first rotational captaincy happened in 2007, and since then I have remained on rotation. This is no bad thing: there is no way I could have maintained a family and seagoing employment with any other arrangement.

INSIGHT

A superyacht has a voracious appetite for your physical and emotional toil. It is emotionless and unrelenting in its demands. Any successful player in this environment will learn the tools to protect themselves.

KEY TAKEAWAYS

- 'Longevity' and 'rotation' are used freely and wrongly in the industry; treat each with care.

HOW TO DIVERSIFY

Speaking of embracing different approaches is to acknowledge and embrace diversity and integration. This is not the 'woke' chapter: it is the selfish-me speaking of seeking the easiest path to leading a strong team/crew on board a superyacht. This awareness came late into my yachting story. It was not that I did not believe that diversity is a good thing, rather that my blind spot towards diversity within yacht crew was so large that I did not even realise I was enabling an out-of-kilter arrangement.

In the early 1990s I was serving in one of the first Royal Australian Navy ships that took women to sea. There were only three women within a crew of 220 men. All of the women were officers, and all were known to me from our shared time in training. They did incredibly well, given that very little compromise was made to meet their needs. There was a slight privacy adjustment to the toilet in the women's cabin, but not much beyond this. We (the existing male crew) were given no training, nor were we expected to amend our behaviour in any way. It must have been very hard for these trailblazers; they were three women among a group of men who had never thought they would be sailing within a mixed crew. There was much talk, both for and against, about this great leap forwards in naval manning and I developed a narrative, original to me, that I liked: 'Any workplace that does not seek to (as close as possible) reflect the wider values and norms of society is destined to be self-limiting.'

'Any workplace that does not seek to (as close as possible) reflect the wider values and norms of society is destined to be self-limiting.'

I may have taken the liberty of curating my own words, but the message remains the same and with this as a base I really should have been prepared to live its truth when I moved to superyachts. In this I failed. I wandered aimlessly in yachting with a narrow definition of diversity and even applauded myself within this constrained frame. I spoke of building crews that were not dominated by any one nationality group and claimed that I recognised the improved performance gained by diversity. Where my narrative fell short was that my frame for diversity did not extend beyond the recruitment of the 'usual suspects': my crews were predominantly English, South African, New Zealand and Australian, with just a few people from other nations – a Scandinavian, maybe a scattering of other Europeans, and in larger yachts this group was supported by some incredibly well-trained Asian crews. I also failed to stretch the frame with gender diversity, with just one or maybe two women within the deck crew.

More than their national, racial and gender similarity, the crew all 'felt the same' – they were cookie-cutter replicates of each other. This was strangling the real benefit of diversity: diversity of viewpoint. Before this Oprah-esque 'Aha' moment, I was missing a beat. I did not actively seek to build a team that featured broad diversity. I just let it happen (or not) organically. I did not look to challenge any existing norms. Due to this, the performance on board was also limited to the 'norm'.

It stands to reason that regardless of any secret sauce of any one crew, the full potential of a yacht, or any professional team, cannot be reached without using all the value that can be created from a truly diverse group.

Edward Herman and Noam Chomsky, the latter being the famed MIT linguistics professor, wrote of the risks of working within a narrow band of dissent in their 1988 book, *Manufacturing Consent*. To be clear, Herman and Chomsky were not writing about yacht crews or any workplace. One of the tenets of this book was that mainstream media promote controversial discourse yet do so within the narrowest bands. We see this when two panelists are brought on to a talk show to vehemently argue their opposing views but do so from very similar socioeconomic positions and within tightly defined boundaries. The book says of this: 'The beauty of the system, however, is that such dissent and inconvenient information are kept within bounds and at the margins, so that while their presence shows that the system is not monolithic, they are not large enough to interfere unduly with the domination of the official agenda.'

I may well be drawing a long bow from the global phenomenon that *Manufacturing Consent* describes and assigning it to yacht crew composition, but like the talk show, I signalled my forward-leaning stance on diversity but did so within the unwritten but tightly maintained boundaries of 'accepted yacht practice'. I write this now as a challenge, to you and to me, to do better. I want to be accountable for my own efforts to shift the needle towards a more truthful diversity in the floating yacht community. In practical terms, when you are

'The beauty of the system, however, is that such dissent and inconvenient information are kept within bounds and at the margins, so that while their presence shows that the system is not monolithic, they are not large enough to interfere unduly with the domination of the official agenda.'

looking to yachts as a new workplace, a new home, you should also be assessing whether the culture will be accepting of views and thoughts that differ.

INSIGHT

No team can reach its full potential without drawing on the widest talent pool possible. Placed in a yachting context, building a diverse crew requires active engagement by the captain and onboard leadership team.

KEY TAKEAWAYS

- Diversity and inclusion matter. Whether you are a captain or a cook, you can help change the culture by actively seeking yachts that promote these values. Everyone benefits from there being a wide range of different people on board.

LEARNING FROM EACH OTHER – *BUNGA TERATAI SATU*

If I am to speak of the performance advantage that diversity delivers, there needs to be some awareness of how diversity is to be understood. To do so, I am going to use the story of a ship that ran aground. Many crew have looked back to me (understandably) completely nonplussed, when I offer '*Bunga Teratai Satu*' as the answer to a situation. Having said my magic words, I expect the crew I am speaking to will have shifted their suspicions from me being mildly eccentric to certifiably bonkers. Let me explain. I'm not mad, I promise. *Bunga Teratai Satu* ran aground in 2001 on the Great Barrier Reef. It missed a scheduled course alteration and ran up on Sudbury Reef, 22 nautical miles south-east of Cairns, Australia, hitting the reef at 23 knots. The ship did the same route every two weeks, and the long-term chief officer and helmsman were both on the bridge at the time. So why did it ground and what does the event have to do with life on superyachts?

Doing complete injustice to the full safety report, the executive summary of the event surmised (I'm paraphrasing here): 'The high-caste Indian chief officer was on the phone when he should have ordered a course alteration.' The Bangladeshi helmsman was at the wheel ready to receive the order, but due to modes of behaviour, instilled in him by the caste system present in Bangladeshi and Indian culture, he did not feel that he could interrupt the Indian chief officer to alert him to the danger. Going aground seemed a less confrontational option. This was an example of a high 'power distance' at play. As described in the book *Innovation, Entrepreneurship, and the Economy in the US, China, and India*:

> Power distance indicates levels of authority in a hierarchical structure. In high power distance cultures, there is a significant amount of inequality between different levels. Those who are lower in the hierarchy are rarely involved in any significant decision-making. In this environment, position power is far more important than intelligence, diligence, or competitiveness, and a person's fate is usually determined by those in positions of authority, rather than by their own efforts or merits.

When my lecturer at maritime college, Rajiv, introduced the term 'power distance cultures' to us and then backed it up with examples from the aviation and maritime industries, so many pieces fell into place for me. As so often happens, it takes someone else to provide the words and the logic for something you have experienced to make sense. Going back to the Great Barrier Reef and the Indian/Bangladeshi bridge team, they were products of deeply hierarchical cultures, and their actions were governed by these rules, rather than the common sense required at the time.

At the other end of the power distance curve are those cultures that tend not to see the difference of rank, education or caste/class. Australia, South Africa and Scandinavian nations all sit within this group. When leading or living among a yacht crew made up of Asian,

European and southern hemisphere crew, it is vital that everyone is seen and included. To achieve this you need to understand where you and your colleagues sit inside the power distance scale. Who are the ones who will not speak out to their superiors and who are those who will challenge all information? Bear in mind that cultural generalisations are just that – generalisations – and the risk with this is that you might make false assumptions and pre-guess a behaviour. I have worked with South Africans who are rank sensitive and Filipino crew who will challenge openly and regularly. So regardless of where people are from, everybody needs to be seen and to feel their contribution is valued.

INSIGHT

Without our conscious awareness, we all seek to be 'seen' by others. We want to have questions asked of us, to have our dreams respected. The fastest way to achieve this for yourself is to do the same to others. Some will share openly and some will need it drawn from them; in all cases it is important to make the effort.

KEY TAKEAWAYS

- We all carry our biases, and our culture and background will impact our behaviour and willingness to speak up in any given situation. Fostering an environment of inclusion that encourages everyone to make their voice heard is very important, and it might just stop you running aground.

LEARNING TO BE MORE THAN CREW

A lot is demanded of yacht crew and captains. The traditional path of entering a yachting workplace is as I have described: moving with time and training through the hierarchy, from new entrant to competent crew, to life as a demi HoD and then to departmental charge or captaincy. To gain the breadth of experience and knowledge that is needed, crew who have an interest in making a career within the superyacht community should consider how they are going to weave in added 'off the yacht' learning.

My own journey with this approach was more by accident than design. After an extended period of working on board I took a break with my family to think things through – career, life and more. During this time, I reflected on where my career might continue. Would I even be considered again for captaincy? I respect my colleagues and knew that there were more good captains entering the industry all the time. Would my skills stay current? It is a fast-moving industry and maybe I would be overtaken if I did not return immediately to command. The choice was not entirely mine to make, as there were no captain positions available immediately.

In lieu of this, a yacht management firm approached me and with the timing and the offer matching my circumstances, I joined their team. In the next 18 months my knowledge accelerated at a rate far beyond what it could have if I had remained on board. I was absorbing information from the many yachts in their fleet; I was seeing the good, the bad and the ugly. I was observing captains 'in the wild' and was able to notice with an objectivity I did not have when I was sailing which behaviours were successful and which were limiting. I was beginning to track what I would take into my own future captaincies – something that would not have been possible without the period ashore. My experience relates to captaincy, but it is wholly transferrable to other departmental pathways.

If I look back on my career, full of peaks and troughs, my greatest knowledge gains have come from the diversions along the way. I may not be the fastest learner, but I would not have been prepared for the challenges of my seagoing career without the experiences I had during time spent ashore. Keeping yachts at sea requires many supporting and co-working entities: suppliers, shipyards, shore agents and brokers. Much is made of the relationships needed to succeed in yachting, but the foundation of any relationship is that there is an understanding, an empathy, for the other person. This need is exposed further when one party is ashore and one is at sea. In my early days, I fell clearly into the frame of the 'hard-done-by' seafaring captain being harassed by the shore manager who had no understanding of the environment, the pressures, the hardships I was experiencing. It came as some surprise, then, when some years later when I was supporting yacht captains from ashore, I was confronted by them being equally full of self-indulgence for their own situation and defensive when I made 'reasonable' requests. The same 'reasonable' requests I had bristled at when I was sitting at the captain's desk on board.

I returned to sea after my stint at the yacht management firm and at times there were days when I slipped back into my natural state of 'hard-working and hard-done-by yacht captain' who was leading a 'hard-working and hard-done-by crew'. I regressed, peppering the shore support office with a constant stream of 'urgent' requests. On good days, though, I retained my awareness of what I had learned during my time in the office: I was more sensitive than before and more respectful of their lives ashore. They were not working seven days per week (on rotation) and had lives and families to balance against their work. I was more critical of my and the heads of department's tone. There was no need to be so demanding all the time and I was appreciative that those ashore held a balanced perspective.

In my first book I wrote of an incident, largely of my own making, where while towing a tender through the night I felt I was at risk of losing the boat (value US$1.5 million). I was tired – it was 3.00am – and I was upset that I'd allowed the situation to develop. Luckily, my experience allowed me the maturity to see I was no longer processing the facts clearly. So I rang my shore support office, gave the facts as I could see them to the shore manager (who in my mind I referred to as 'my conscience ashore') and asked for their perspective. In the specific incident, the non-seafaring yacht manager could not give me the silver bullet to fix the situation – it did not exist. However, he was able to listen, reflect and give me the support I needed. I would not have thought to do this if I had not spent the time prior ashore and developed my willingness to draw on the expertise and support of others.

THE VALUE OF REALLY SEEING PEOPLE – THE DIGRESSION

Having just spoken of the importance of professional digression to yachting success I am going to lean on this and speak of another digression. Digression being a delightful euphemism used by Liz, my wonderful editor at Bloomsbury during the writing of my first book *Superyacht Captain*. Liz's comments often began with, 'Brendan, do you think this is a digression?' The question was rhetorical at best – they were all digressions. I defended some, but Liz was correct: my meandering stories did not add value to the message of the book, and while each word was a friend, I cut without fear or favour. From the 30,000 words that fell to Liz's deftly wielded axe there was only one adjunct to a story that I am taking the opportunity now to re-craft and re-submit. It is a story that is at the heart of life as a (mostly) mature superyacht crew member.

In that book there is a chapter called 'You can have the shoes' in which I spoke of my closest childhood friend, Michael. He was a boy whose family were just a little wealthier than our town average and had a view of the world just that little broader. While writing the first draft, I racked my memory to commit to print what they had really shared with me. They didn't particularly promote my education, challenge me to travel or give me any specific guidance, so what did they do for me that means I see them as potentially the greatest single influence in my development? I didn't ruminate enough to find an answer in *Superyacht Captain* but, as luck would have it, I have chanced upon it since.

I returned to Western Australia for a few months after publishing the book to do some interviews and used the chance to catch up with some childhood friends. I saw Michael and his family and within the first ten minutes I had the answer that had previously eluded me. It was twofold and spread across linked spheres of influence. The first was that they shared their wealth with me through the long summers of my teen years.

The term 'wealth' needs some deconstruction here, as it is too easy to define it as a financial measure. This is not what I mean. Although the family must have paid for me during the many shared activities of our youth, and they fed me incredibly well. When I speak of wealth, I am referring more to their wealth of experience, friends and conversation. They included me in everything as equally as their own children. I heard risqué stories and laughed at bawdy humour not normally made available to children, and observed behaviours that the parents manifested from their lives well-lived, which would stay with me into my own future.

The second and more important answer I had to 'why the life-changing impact?' was that Michael and his family were the first people I can recall who 'saw' me. To 'see' someone is possibly the most wonderful gift you can share with someone at any age, but to truly see a child as Michael, his brother, sisters and parents did me, was life-changing (thank you eternally, family W).

Until I joined their unruly and fabulous table I had never been asked for my input. The now outdated mantra that children should be 'seen and not heard' was a driving force of my own family. I had never been asked for my opinions, my observations or to share my hopes and dreams for the future. My parents, grandparents, wider adult relatives and their friends all reflected the mores of the time and could not see that children had, and could, add any value by their input. The adults of my experience would play games with children – indeed some of my best familial memories are of backyard cricket and street tennis – but they would never consider that anything a child would say could be worth listening to. This was in such stark contrast to Michael's family's approach, where the conversation was the point, and all were to take part. Hopes and dreams are the greatest currency a young child has and to finally be asked to share mine with no judgement being passed was to be truly seen and then in turn be empowered to follow them.

I am now a parent myself and while I do not think I can replicate the breadth of conversation I experienced with Michael's family with my daughters (I am too prudishly self-conscious to share the messy realities of adult life), within my limitations I do take the time to listen, reflect, question further and most of all respect my daughters' contributions. In doing so, not only am I modelling a behaviour they may take through their own lives, but I have learned much more about my children: one has exceptional comic timing and the other has a way of seeing many of life's problems through a very different prism.

This childhood story may not seem a direct link to life on board superyachts or any future career, and Liz's first assessment may still seem valid, but I argue that it is critical. Yacht crews are multicultural and multigenerational. Leading or even working inside such teams requires emotional maturity, and 'seeing' people for their contributions is a fundamental facet of living a great superyacht life. I was fortunate in my studies that I chanced upon a maritime lecturer who focused less on the theory of ship construction and more on the human factors on board a yacht. Rajiv provided the words, the studies that backed up what I thought I knew. He spoke of cultural awareness, stress and the importance of trying to see your shipmates clearly. My story of Michael's and my own families was a top-down narrative of parents responding to a child, but seeing people fully and truthfully works in both directions.

Key takeaways

- 'It is lonely at the top' is a cliché because it is true. When looking to the captain or any leader, understand and remember they want to be seen as a person first, just as you do.
- It is very easy to become so absorbed by the day-to-day business of life on board a yacht that you forget to take the time to sit back and try to see your colleagues for the humans they are. If you can manage to do this, and make that person feel truly 'seen', everyone will benefit.

LEARNING PERFECTION – THE HARDEST TASKMASTER

There are many competing pressures in the superyacht environment: the unforgiving sea, the equally unforgiving yacht owners whom we met earlier, the challenge of performing in your assigned role under scrutiny, and the importance of learning the cultural communication paths needed to work within a diverse yacht crew. These are all very much in play, all day, every day, but there is another unseen yet tougher guiding hand that influences the entire culture of superyachts. I was already living within it when a non-yachting friend said it out loud.

In 2003, the America's Cup was being fought on the Hauraki Gulf, the windswept waters adjacent to Auckland, New Zealand. The yacht I was employed within was the 'home away from home' for one of the team owners. We had relocated the yacht from Europe, via Florida, the Galápagos Islands and Tahiti – yes, sometimes the work is pretty great. The yacht's owner was gracious and the captain at the time gregarious. The combination meant we had the opportunity to share this environment with friends and family on an 'open day' watching the America's Cup racing. This was a once-in-a-lifetime opportunity for non-billionaires and all invitees were suitably in awe of the environment they were sharing. Yvonne and I had two very close friends in Auckland – friends who had moved from Australia to New Zealand to further their medical careers and because the windsurfing was fabulous. They were having a great day, as were we, and it was a pleasure to show them around. The couple had a keen eye for design, and they were so appreciative of the opportunity to see the splendid Japanese-inspired interior.

We were doing the tour: the tenders, the jet skis, the wine room, the gym and the engine room were all shown. By convention, the guest areas are not open for non-crew and non-guests, and rightly so – these spaces are someone's home and should remain private. I was enjoying my friends' reflected admiration of my workplace, but they had to leave to go to their respective shifts at the hospital, so a tender had been arranged for their transfer. To do this the yacht slowed a little and changed heading to provide calm water for the tender to approach. The bridge called to confirm all was OK and with my friends behind me, I opened the hydraulically operated side door of the yacht. This exposed a section of damaged wood and a rusty frame that was concealed when the door was closed, as it was 99 per cent of the time. The 'wound' caught the eye of my friend, and as he ran his hand along the damaged wood and the rusted surface he said, 'Perfection, eh? A tireless taskmaster.' It was spoken in a gentle and reflective tone, not to convey any disappointment or criticism, but rather as a philosophical observation.

In the years since this moment, I have reflected on just how perfectly his words capture the contradiction, tension and irony of yachting. The constant goal is to maintain and deliver perfection in an environment that is perpetually working against this outcome. The delivery of perfection is a false and impossible mountain to climb, since the conditions and the frailty of humans are all working against it. Yet nobody in the superyacht community, ashore or afloat, wishes to say the unsaid out loud: chasing perfection is a pointless pursuit. For the paint provider, for the linen supplier and for the carpentry contractor it would be a commercially foolish thing to do. For crew, it would be an admission that human error occurs on board, as it does in every other sphere – that it does not stop at the threshold of a superyacht. And yet having an awareness of the almost comical impossibility of achieving absolute perfection all of the time is important. It acknowledges our own humanity, and that's an important thing when you're all at sea.

Key takeaways

- Although it's something everybody strives for, perfection is actually impossible. Try to remember this and treat yourself and others with grace if something falls short.

MAKING PEACE WITH FUTILITY

If perfection is an emotionless and demanding taskmaster, then futility is the jester in the court. In a time when superyachts are scorned for being a symbol of wealth inequality and derided for their outsized environmental impact, I am often asked to defend my work and by extension the wider superyacht industry. I say quickly, it is indefensible, but this is a deep wormhole to fall into. Activities that squarely sit alongside yachting in the indefensible camp include, off the top of my head, snow sports, stadium concerts, motorcar racing, low-cost holiday travel, leaf blowers and pet ownership. All are integral to the lives of many and lift the human spirit, but they are not required and from an environmental perspective, they are generally indefensible. This is intentionally reductionist to the point of inviting ridicule, but my point is to caution against shaming others, as we all share the stain of environmental guilt for our discretionary choices.

To demonstrate the futility of seeking perfection, I'm going to tell you a story about cleaning windows. Stay with me. It was warm and humid in Savannah, Georgia, USA and with light winds the sulphurous odours from the paper mill drifted along the Savannah River. For a tourist destination with an exquisite town centre, the presence of the paper mill always left me a little miffed. Surely there could be a better location? Town planning aside, the 80-metre yacht we had brought into the shipyard some months before was languishing; the captain was in a battle with the shipyard as much as he was in a battle with his own demons. He was choleric and erratic, and my role as the first officer was to keep things moving along without any direct interaction with or contradiction from him.

One of the projects was to improve the access for the deck crew to clean windows from 'over the side'. This refers to the necessary, though dangerous, practice of suspending yacht crew in harnesses outside the yacht. By dangerous, I mean there have been many serious accidents and even deaths from this work practice. To conduct the same work in a shoreside setting requires licences, certified training and certified equipment. Now, this is also (mostly) the case in yachting, but this was 2006 and safety was still making its way from being 'a nice idea' to a core value. We were too focused on getting the job done to pause and acknowledge the risk we were exposing ourselves to.

Crew are not lifting equipment professionals, and we were not trying to overstep our role. The plan was to make a relatively cheap mock-up of the yacht to pass to a professional so they could build the certified solution. The goal was to allow crew to be safely suspended over the side of the yacht at heights up to 10 metres above the water or the dock. Making the mock-up was where we started the work, but there had been quite a few runs to the hardware store, and we had all committed too many hours to count. The simple solution had lost its simplicity and if our mock-up was an indication of what was required in reality, then the permanent solution would entail a six-figure sum to ostensibly clean a couple of windows. Somewhere in the heat, working in shipping containers, a bosun said, 'I hope this isn't futile.' An ever-quick-witted English second officer paused, caught the attention of the group, then replied, 'This work may or may not prove to be of value, but the entire industry is futile. We add no value to the greater human endeavour. We build nothing. We create nothing of value. The only way to remain working on superyachts is to be at peace with this.'

When viewed on the page these words seem harsh, but they were not intended to be. The second officer is a storyteller, and his observation was made with the goal of injecting humour into our day. Nevertheless, this light comment remained with me, and I have spent a lot of years measuring my days' worth by the happiness, or not, of a billionaire guest. On a good day I do not allow this to affect my own sense of self-worth, but when I am not so strong I draw on the pithy observation: my work has a sense of futility, but I am at peace with it. To look for a higher value would lead me to a torrid emotional beating.

KEY TAKEAWAYS

- Yes, superyachts are surplus to the world's requirements. Yes, they are a futile endeavour, but so much of the joy contained in the human spirit relies on equally futile activities. To deconstruct joy to its virtuous components is to risk removing it entirely. In order to succeed in superyachts, you need to recognise and make peace with this.

BUDGETS AND EXPENSES – NOT THE MAGIC PUDDING

Talking about perfection in yachting inevitably leads on to cost. Yacht ownership is an expensive and complicated game. Some of the larger yachts exceed warships in their size and they need to be funded. It might be a fair assumption to expect that running yachts with asset values in the hundreds of millions and being tasked with delivering 'unbelievable experiences' means that there must be bottomless budget behind it all, a magic pudding that returns to whole each time a slice is taken. This assumption could not be further from reality.

Yachting expenditures are forecast, monitored, scrutinised, argued over and, in my experience, always reduced. It is a nod to the weight of expense that a yacht creates that even among the heftiest billionaires the burden is felt. I try and scale it down and use analogies to make sense of it when talking to people outside of yachting. For example, maybe you can afford the purchase of an expensive car, but the insurance, service and ongoing costs will still make you wince and question the purchase. Maybe so much so that you challenge the dealer who sold you the car and the service centre that maintains it, demanding to know why the costs are so high. If you magnify this by many multiples, then we are nearing the sense a large yacht owner has when the operating and maintenance costs become visible.

A new entrant might feel themselves surrounded by yachting extravagance, to a point where nothing matters, and where waste and excess are baked in. This may be a fair first impression, but it is not the case. The yacht owner does not accept that excess and waste are compatible with yacht ownership, and they are always on guard for those they sense are taking advantage of their wealth. The people who have responsibility for expenditure on behalf of yacht owners, their CFOs, their trusted inner circles, also take their work seriously. Do not be fooled into thinking financial frivolity or waste will be tolerated: the owners of yachts are in their position due to an acute sense of cost and value.

Of a hundred different examples I could have drawn upon, I'll use one that is quite recent. The yacht was alongside in Monaco and over the period of several days the stores needed for an ocean crossing and an extended period in the Indian Ocean were being loaded. It is rare to load stores with the owners on board, but the schedule was tight and, being experienced owners, they knew what had to be done. The principal was returning from a walk as the crew were loading, and I stepped out to greet him. He was perplexed that during all the time we were talking the only thing that came on board was frozen fresh milk. He estimated it must have been 500 containers and his puzzlement had an edge: how much had all this cost? He then asked the chief stewardess to look up the price of milk in France at the supermarket Carrefour. The challenge was that the price paid by the yacht, having bought the milk through a yacht support agency, must approximate the retail price. For a moment or two I was thrown into turmoil, as I was not sure this was the case. Within a minute, the chief stewardess returned with some figures. The price difference was only 3 cents per carton, in favour of Carrefour, who could not have delivered in the volume needed. We survived this challenge, but the point was made: do not take the owner and their wealth for granted.

An extension of checking every purchase with rigour is developing operating budgets to forecast such expenses. Given that annual budgets can run to tens of millions it is not something to be taken lightly. Fortunately, yachts' expenses are repetitive and largely foreseeable. The temptation is to sidestep detail by using the excuse, 'It varies depending on the owner's choices.' This is a trap, as with the correct modelling, all variations can be forecast quicker than the expense can be committed to.

The big variable drivers are fuel, remote area expenses and large changes to insurance in the case of war risk areas (e.g. transit of the Red Sea). The calendar is the multiplier and the variable in the equation. Using this variable, most costs default to unit cost × consumption of units per day × number of days. Where time is not the multiplier, there will be another number to justify. Working alongside a very organised chief stewardess I should have known better than to dismiss her request for a purchase as under-researched. I was caught by looking at only the total: €9000-plus for coat hangers felt high, given I would baulk at €10 for the same items in my own life. Her logic was sound, and she did not pause when defending the expense: 15 guest cabins × 50 hangers per cabin × €12.50 per hanger (type chosen by the owner). This granularity is needed across every line item, from fuel to laundry liquid. When I now develop budgets, I still look at every number as though it were my own purchase. To lose track of this perspective is a road to ruin in yachting.

Much of this budget-setting information will not directly apply to you if you are in your early days, but it is important to know how budgets are drawn up and why margins are so tight in such a luxurious setting. I still feel, as I did when boarding in Monaco on my very first day, a degree of amused confusion that one person can amass so much. I still move around the yacht looking at the minor items – the uniforms, the plates, the cleaning cupboard – and marvel that it all belongs to one person or one family. It's no use justifying in your mind a salary increase by looking at the fuel spend. This counting of apples as oranges does not work and does not lead to any productive outcome. Money is spent on different things in different ways, and everything is accounted for, including you.

Key takeaways

- Superyachts are incredibly expensive to run and maintain. For this reason, all expenditure is very tightly controlled and everything is budgeted. Do not think you can take advantage of the wealthy owner or be frivolous or wasteful while on board.
- It's also no use carrying around a sense of entitlement when it comes to pay. Salaries are very carefully calculated, so comparing your monthly pay cheque to the cost of some cushions or the fuel bill is futile.

PART THREE

LEADING ON BOARD

Part 2 was all about learning to set yourself on a path to success in superyachts. In Part 3, we will move on with some reflective learning derived from my own captaincies. I do not profess to tell the total or only story of yachting or leadership, nor to extrapolate my lived truths as truths applicable across all other fields and lives. Nothing I share should be read as irrefutable and all learnings are open to challenge. That said, I have been able to draw some conclusions from my experiences, which should be useful for anyone entering into the superyacht sphere.

6

LIFE AS A SUPERYACHT CAPTAIN

HOW TO IDENTIFY AND USE COGNITIVE BIASES

As outlined earlier, my career has meandered through the Royal Australian Navy to commercial shipping and settled with superyachts. Learning from all these environments has added value to my life and, I believe, fed into how I function as a captain on board a yacht. These experiences have also informed my embedded cognitive biases, which even though I recognise them, I cannot shake. Like all of us, I am not always rational. I can only say out loud where I think my biases and irrationality show up, thereby acknowledging them publicly. Doing so forms an accountability pact with my colleagues and crew and gives them a chance to challenge them when they inevitably surface.

I also continue to question my past influences and try to understand how they impact my leadership voice. I had an austere childhood where initiative was not praised – it was expected. Then, a period in a defence institution followed by a life spent travelling meant that I lacked a cultural home, which in turn engendered a certain insecurity. I can feel both the positive and negative pull of these experiences and their effects when I am faced with a challenging situation. I may view others as indulgent in their requests, though I ignore my own indulgences. I am short with those seeking assistance, yet I am scornful of those who do not provide the support I need. And I may take a self-righteous position when people are parochial about their local area, but internally I envy their sense of connection to a place. On the flip side, there are positives to my biases: I can adapt quickly to new environments, I am open to different views and I am the opposite of fragile in the face of adversity.

If the event warrants, I seek to balance these biases (for better or worse) through reverse-engineering a leadership or decision path, by finding the academic study to support my conclusions about the event after it has happened. I believe this approach to be an authentic, though sometimes painful, way of exploring leadership and decision-making.

This process of evaluating and learning from past events has been at work throughout this book, and it is something I am proud of. I understand that this serves my ego, but when I listen to some clever and professionally polished speakers, podcasters and authors speaking about leadership, the sceptic within me wants to ask, 'What real leadership experience have you had?' When they talk about the influence of their high school sports coach as their defining leadership experience, they lose me. How would they perform during a fire at sea, a life-threatening illness affecting one of their team mid-ocean or the death of a co-worker, with the attendant task of informing the parents and loved ones? Having been through all these, I stand by my conviction that my life's investment in researching and learning leadership lessons by reverse-engineering leadership theory gives the value in this book's message.

To place my experience against the high-distribution, highly polished 'retail' leadership advocates may present as arrogant. I hope it isn't, and I say that with a sense of introspection, but my toolbox has grown deep through the years. The experience in the military, commercial shipping and in superyachts has provided more cross-over in terms of leadership skills than I first thought was possible, and a cumulative leverage that each in isolation could not have delivered.

Leadership and the training of leadership have long been a core business for the Navy. They know that ships are platforms to move weapons and that this can only be done by well-trained, well-led teams. I learned this lesson well during my time there. Next came commercial shipping. The commercial waterfront has been stripped bare for many decades. What I mean by this is that 'the do more with less' brigade that moved into the wider corporate world in the 1970s and 1980s had been at work in shipping a decade earlier. Leading complex operations on an ever-shrinking budget creates a ratchet-like pressure. Overlay the non-reversable consequences, where errors can and do cost lives or catastrophic damage to the environment, and working in that environment becomes very real, very fast.

Superyachts, on the other hand, present a unique leadership challenge, drawn from a bottomless set of expectations and a lightly regulated environment where the 'norms' that place the lines on the leadership court are not visible. The freedom afforded to a yacht captain at sea could be seen as the greatest privilege, and it is, but it comes with a balancing burden of flying without the safety net that is present in a shore corporate environment that is regulated by domestic labour laws. Almost all yachts sail under offshore registration that upholds safety, but allows a much more permissible view of labour protection. This is not cause for alarm – but you should be aware that you are flying without many of the domestic labour protections.

KEY TAKEAWAYS

- Pause now and have a think about what the greatest biases are that you bring into your life and your decision-making. Write down the three to five biggest influences that drive these. Are they familial? Cultural? Societal? Economic? Educational? Professional? Geographical? This isn't an exact science, and you should revisit these often, so don't worry too much about it. It's just an interesting exercise that will help you to name your biases, both to yourself and out loud to those you lead.

- There are countless books, podcasts, TED talks, articles and videos on leadership out there. Explore these and see which resonate with you (or don't!) and write down any takeaways that you can apply to your own behaviour – past, present and future. It is also worth taking the time to sit down and reflect on your own experiences and see if there are lessons you can draw from them.

WHY LISTEN TO ANY CAPTAIN?

This is a very good question, and one that my crews over decades have also sought the answer to. It would also be a fair extrapolation to surmise that this question arises in crew messes around the global superyacht fleet. I do know that in my story, some crew did not listen, some listened in passing and a small group would remember and speak my words back to me some unfathomably long time later. Sometimes, this would be to point out an inconsistency, a hypocrisy that I was dimly aware of, but hoped nobody else would spot.

I was very proud when I became a captain – it was slightly surprising for me – and I responded by expecting everyone else to view this part of me with the same awe. They didn't of course, and my pride was mistaken for arrogance. It was a Damascene moment in humility to realise that what I viewed as a professional and life highlight was viewed externally as 'just another captain full of his own importance' and that I was really a very small part in a far bigger operation.

My moment of realisation speaks to the fact that captaincy is a wonderful place in which to find out some truths about yourself and those you lead. There are many forums in life and business where leadership is essential, but there are few, short of a battlefield, where there is no chance to take a step back from the line of fire. This relentless spotlight effect has broken many. The demand that you be the leader that you want to be for your team, that your crew need you to be and that the yacht owner expects you to be, can be crushing. Business books and leadership mantras often only work in an environment where there is the chance to put your 'game face' on for the hours spent in the workplace. Those people have a chance to psych up each morning for the rigours of the day and to decompress each evening in the privacy and comfort of home with those closest to them. The texts cover how to present in meetings, how to lead projects and how to give professional feedback. They stay inside the professional boundaries that the shore-based workplace defines.

These boundaries are a luxury not afforded to captains, or really any yacht crew. In this environment, if you are not true to yourself then the fatigue of maintaining a false narrative for weeks or months in a confined environment will break even the strongest. Conventional management education does not give the answers to – or even ask – the question of how to live adjacent to your team, where not only do you work together, but you also dine together, socialise together and live in each other's pockets. Even the greatest, most charismatic CEO does not carry the food, cleaning products and toilet paper into the office alongside their co-workers and have their cabin inspected for hygiene.

KEY TAKEAWAYS

- Be authentic – if not for your crew and your employer, then for yourself. Too many start strong, projecting the leader that they think they should be and not the one they are. This will fatigue even the strongest.

PUT YOUR OWN OXYGEN MASK ON FIRST – HOW TO LOOK AFTER YOURSELF

This is a tough section to attempt, mainly because it has been so widely discussed and elicits such varied responses from different 'experts'. So long as YouTube has like and subscribe buttons, self-awareness, self-help and self-indulgence guidance will continue.

My own approach, as clarified earlier, is to look at events I have experienced and play them back later to see what was going on at the time. The truth is that I am rarely aware enough in the moment to know what is happening around me, and it is only with reflection that the real learning happens.

From doing this, I understand that I, in earnest, view the leader's role as being to provide a service for others: the servant leader model. This conforms with the view of many TED-talking, podcasting, social media influencers in the field of leadership. It plays well to the room to speak of the modern leader (i.e. the influencer themselves) as someone who does not self-indulge, and to explain how they view their performance as a measure of the output of others. They are humble, compassionate and can craft an amusing anecdote about themselves as the contrarian, flawed performer who always puts the team first. However, to say 'leaders eat last' is very different to actually being the one eating last.

I know this for a fact, because I tested this approach, with the absolute best intentions of putting others first. My leadership faltered to failure. I am warming up to tell this cautionary tale by breaking it down into components and providing some frameworks to facilitate this painful retelling of how I became a better leader. Please bear in mind that although I am writing this from my perspective as the captain, it is equally instructional to all crew affected by the captain's performance.

From pleasing to passing out

I probably sit somewhere just to the left of centre inside the bell curve of the professional captain diaspora. I am an able operator, but I know there are better. I am a growing communicator, but when I watch a natural at work, my inadequacies are visible and in a social setting I can veer to the awkward. Even though I can name these traits and write them here, I recognise that they do not define me; no leader can be defined by just a handful of attributes. Captains are drawn from many cultures, with a huge diversity of backgrounds and views. This is good in many ways, but it also brings with it an oversized grab bag's worth of biases, both conscious and unconscious, into our leadership practice. Captains agree and disagree in equal measure, and we can be prickly with each other, lest the shortcomings of our peculiarities be exposed by a peer. We can be self-serving in our humble-brags of seamanship achievements and we are always defensive and protective of our crew. We may talk them down ourselves for their performance, but heaven forbid if another joins in this chorus.

My assumptions above are safe, based on personal observation, but until recently, they were poorly researched in the field. For all my time sitting inside and defining myself within this cohort I do not know many other captains with any intimacy of connection. This is simply because, thanks to the very nature of our role, we do not spend very much time in each other's company. The moats around our offices are not virtual, they are real. Modern business parlance cautions against falling into 'silo thinking', urging leaders to avoid making decisions in a vacuum and building walls. Yet the walls separating yachts are enforced both structurally and geographically and only an unusual circumstance can peel them away.

One such circumstance presented itself to me during the writing of this book. I agreed to assist with a course that saw 12 to 14 superyacht captains meet weekly via video conference for an eight-hour training session. The purpose was to deliver mental health awareness training, and my role was to give context to the captains and to help the trainers by prompting for engagement when the course members were too reticent. In truth, my role was largely to nod in the background as the professional course facilitators delivered quality education.

Alongside the learning itself, this opportunity provided me with eight hours a week to watch, listen to and learn intently from more superyacht captains than I had ever previously met in my career. Some I knew by name, some I had corresponded with in the past, but I had not spent any meaningful time with them. I accept that this situation is not a unique attribute of captains

– many industries operate remotely, even if that remoteness is just a separate floor in the same building. But add to the mix the fact that most peers are commercial competitors, and it creates an environment ill-suited to openly sharing life's struggles – personal and professional. To overcome this, I sought to find a thread, a chain, something that bound the other captains and me together by shared outlook, and something I could hang other ideas from.

With a gentle nod to Sigmund Freud, I thought perhaps that something from childhood might bind the group. Maybe the captains were constrained in early childhood by a 'rules-based order' of domesticity that they wished to sail away from? Or in their teens, when they joined the yachting community, maybe the captains sought a life of adventure on the sea. If either of these was correct, then they belonged in the distant past for the people on the course, who were several decades past first youth.

Some captains on these calls remained evangelists for their profession, some were realists, and for many/most, the gloss had faded; they saw the world in muted tones due to the ravages of time spent in captaincy. On my good days I am truly the evangelist, though I have moved across the spectrum when dealing with the logarithmic curve of billionaire yacht owner expectation. During the weekly video calls, the captains set out their career highlights and dominating these were tales of great ship-handling and carefully curated stories of exceptional delivery of service to yacht-owning guests. These captains liked to, and needed to, please others. This validated my own behaviours, but as with my own experiences, this seeking to please usually had a darker side.

The captains consistently told stories of wanting so hard to please the yacht owners and their guests that they put all other considerations behind this one goal. 'Considerations' here is a euphemism for their marriages, their own health, their children and the teams they led. Each week there was a new group, and although each group was different to the one before, they had in common that someone would share a new story about the collapse of one or all of their 'life pillars'. In one call, a captain retold a recent conversation with his yacht owner. When he was asked, 'How committed are you to this yacht?' he had replied with words that silenced the group: 'Well, I have already given it my health, my sobriety and my marriage.'

It was not uncommon during the sessions for a captain to share with shortening breath their story of taking an extended time away from their yacht to recover their mental and physical health. They never spelled it out that they had suffered a breakdown, but everyone on the call knew what had occurred. This was just one cohort of yacht captains, but while the stories might change, the behaviours and the resultant impact on lives and families is consistent across many high-performance, high-pressure environments.

INSIGHT

As my first yacht captain would say (read in a strong Afrikaans accent): 'My boy, you must give some of your soul to make it in this industry, but only you can know how much you are willing to give... Otherwise the yacht will just keep taking.'

KEY TAKEAWAYS

- To be an effective captain, you need to look after yourself first. This might seem selfish, but you must help yourself in order to help others.
- To accelerate your professional ascent, it is tempting to give your all – your time, your health, your morals. Resist this impulse. Know, write down and commit to your deepest memory what parts of your life are off-limits. Protect these; they are your inner chamber. To not do this will precipitate an earlier and more rapid descent from whatever height you may achieve.

HOW TO SAY NO AND SURVIVE

All of this is taking us into a dark place too quickly and forgetting that while our professional environments are demanding, there is a chance to control them. In one memorable call a highly regarded captain, and friend of mine, spoke openly about breaking down and weeping in front of the yacht owners he was employed by. He had given so much of himself that his wife had left, he had lost contact with his daughter and he found his life was no longer in his control. The outcome was uplifting: his employers were understanding, and the summary was, 'Take the time and make the changes you need to return as your best self.' It took months and some structural changes to his employment and his life, but he returned stronger and better than before. It continues to bring me pleasure to see photos of him hiking with his daughter. He caught his slide before it was too late.

I reflected during the calls with the other captains on my own journey and realised I too was afflicted with this need to please. It is intoxicating when a billionaire lavishes praise upon you for your efforts – when your creativity, matched by experience and skill, delivers the unimaginable – but it is a poisoned relationship. Next time, you must be *more* creative and the experience *more* memorable; the expectation ratchet only works one way. Trying to deal with this resulted in a downward spiral for me and for the team I led. The lines between realistic, challenging and ridiculous became blurred as I sought praise. I could not see the toll this was taking in real time myself, and because it was incremental, the crew I led only saw it some days and not others. In fact, it was the yacht owner whose visits were at longer intervals who clearly recognised my decline. I am sharing the story below as it played out in my environment. Your location may vary, but the underlying message is universal:

It was a beach barbecue on a remote atoll in the Indian Ocean. The location was so splendid that the beach event had to match. There was a DJ booth built from coconuts (true, I have the photo), more watersports than are reasonable to list, games for the children, a cocktail bar, two chefs and an expansive menu. Given this all had to be shuttled from the yacht on the small tenders, it was more beachhead invasion than beach barbecue. Early in the afternoon, adjacent to a volleyball court set for the day, the yacht owner called me over to join him and his wife. I moved quickly over to them, noting all the preparations as I passed. I was pleased with and proud of the crew's efforts. There were guests kitesurfing, children being towed on inflatables, and the elder guests were playing bocce. Yes, I was the pleaser captain, and I was delivering. I was ready to receive my praise.

The ever-insightful owner, cold beer in hand, offered a toast to the day. I too was indulging in my one beer of the afternoon, so he, his wife and I touched bottles/glasses. After a small sip, his wife spoke: 'Brendan, you have lost weight. Do you know how much?' I didn't know. I was a fit, healthy 40-year-old and paid no specific heed to my weight. She went on: 'You look gaunt, Brendan.'

I was waiting for my praise and felt that if these personal questions were the path to receiving this, then it just showed they cared. The husband then asked, 'Are you sleeping well?' I said that when the yacht is operational my sleep is interrupted, and I am used to this. Even though these were my words, they rang very hollow. Our conversation drifted back to the children, their children, playing and enjoying the time. I excused myself and returned to organising something, anything, and then without seeming to hurry, returned by tender to the yacht. Without drawing any attention, I poured my beer away. It no longer tasted very good. I was not going to gain the praise I sought, and the conversation had left me feeling a little uneasy.

Fast-forwarding some weeks, the yacht owner and I argued about a subject of no significance, to the point where there was no recovery. I departed the yacht, and my captaincy was ended by mutual agreement. Those are the facts, but that is to gloss over the reality. The reality is there is no mutual agreement with a billionaire. My decline had been observed that afternoon on the beach and he knew I had wound myself so tight that there was no way back. It was crushing at the time, and for a long time after, but in fact the yacht owner had done me a great favour. He had recognised the tension I had built into the relationship and what it was doing to me, the crew and ultimately his family, who relied upon me for not only their enjoyment, but their safety.

There followed an appropriate period of months during which the yacht owner and I were not in contact, but at some point I felt the need to 'settle' and I made the effort to write a note thanking the family for all that they had given me, by way of the employment, but more importantly, for the chance to grow through the realisation I needed to place boundaries in any future workplace. I was surprised by the detail and effort that went into the response; the yacht owners really did care, and the words were very supportive. We continue to maintain irregular contact and I remain ever grateful for their influence on my life.

Returning from my reminiscence and jumping back to the weekly call: on each captain's course there would also be an outlier, a sage survivor in the hamster wheel of yacht captaincy. These people were often a few years older than the rest of the group and their message was clear: they saw the environment for what it was and had learned how to say no. They knew that the sprint for adulation that so many of their peers were chasing was a very short-lived goal. They had understood that consistency was a more sustainable goal than an upwards spiral of failed promises. Their wise confidence was exasperating to the rest of us struggling to manage the demands of our days. In a moment of true frustration, one budding captain burst out saying what we were all thinking: 'You sound so sure of yourself, but in our environment, how do you say no and survive?' The response was so concise and perfect that I have used it across many applications ever since. He said that he told owners: 'Sir/Madam, it is absolutely not in my best interest to ever say no to you, so please understand, if I am saying no, it is for very good reason.'

'Sir/Madam, it is absolutely not in my best interest to ever say no to you, so please understand, if I am saying no, it is for very good reason.'

This sentence is both simple and transferable across many endeavours. Commercial tension is the cornerstone of business; it is the arbitrage where all deals are made and it pushes managers, leaders and founders into dark places where saying 'No' cannot be separated from failure. If you can turn this around from a potential point of failure to a point of success, then the client relationship will not just improve, it will do so at scale.

KEY TAKEAWAY

- Successful managers, leaders, founders and captains do not seek adulation: they seek respect by doing what is achievable and defending what is not with evidential, fact-based reasoning.

Guest anecdote: Karine Rayson – the crew coach

Karine has long been a force for change in the yachting community. Her energy and passion for the people of the yachting community is really without peer. Her business entity 'The Crew Coach' does not do full justice (I think) to all she delivers. She holds the yachting community to moral account for their actions and gives a safe place for so many crews. That I can call Karine a friend alongside this professional respect is something I am very grateful for.

Having worked in both superyachting and maximum-security prisons in Australia, the parallels between these two seemingly disparate worlds are striking. The phrase 'golden handcuffs' aptly captures the shared experience of losing autonomy within rigid, insular systems. In superyachting, crew members are on-call 24/7, working relentless hours without the freedom to leave or control their schedules, much like inmates confined by the dictated routines of prison life. Both environments foster a sense of institutionalisation, where personal autonomy is sacrificed for the demands of the setting.

The structural similarities are equally uncanny. Superyachting operates under a strict chain of command, with captains holding ultimate authority, mirroring the rigid hierarchies of a prison's management. In both scenarios, individuals must navigate complex power dynamics to maintain harmony and avoid conflict. The tight quarters and lack of privacy in each environment only heighten the interpersonal challenges. While superyachting offers financial rewards and luxurious surroundings, the psychological toll it can take – burnout, isolation and loss of identity – can parallel the weight of confinement. Beneath the glamour, both worlds demand a price, one often paid in silence.

This comparison becomes even more thought-provoking when considering how each environment addresses skills training and mental health. At Eastern state prisons, such as Ravenhall Prison and the CDTCC, comprehensive strategies are implemented to address these areas, often surpassing the efforts seen in yachting. The Department of Justice's strategic goal to reduce reoffending drives these initiatives, embedding protective and preventative interventions into daily operations. Programmes focused on skill-building, rehabilitation and emotional well-being are not merely add-ons but integral to the system, offering a stark contrast to the reactive and often inconsistent approaches within the superyacht sector. While both industries demand resilience and adaptability, corrections systems proactively equip individuals to succeed post-release – an approach the superyacht industry would do well to emulate for the benefit of its crew.

Having managed numerous crises at sea – including suicide, at risk of suicide, hostage situations and freak accident fatalities – a non-negotiable budget must be allocated for crew mental health and skills development to ensure their resilience, safety and ability to respond to the daily demands of the industry.

The responsibility for addressing these challenges lies across multiple stakeholders, from yacht owners and management companies to captains and crews. The power of influence here does need to come from the top. Honest communication needs to be funnelled to the owner on what is required to run their asset optimally. Furthermore, industry-wide, there must be a shift from reactive approaches to proactive, structured systems akin to those in corrections – embedding mental health education and comprehensive leadership training into the fabric of crew operations.

A small investment in human personnel (crews) will yield substantial returns, ensuring smoother operations, enhanced guest experiences and long-term crew retention. By prioritising mental health support and professional development, owners protect their investment and create a more motivated and resilient team, reducing costly turnover and mitigating operational risks.

Unsurprisingly, Karine's contribution delivers a powerful call to action for the superyacht community. While our care for fellow crew and our awareness of mental health needs have improved significantly, there is still a long way to go.

HOW TO COMMUNICATE EFFECTIVELY

This book is not written 'by a captain, for a captain', but I am using 'the captain' as an avatar of a leader and whether you are a leader today or being led, the messages are completely transferrable. Even if you are in your first week ever on board a yacht – you are the greenest of greenies, or as my great mate Will Martin might say, 'You are as green as a seasick frog on a billiard table' – you will be exposed to the leadership behaviours of your captain (keep the baboons front of mind). Their performance and your ability to understand what they are and are not achieving is critical to your enjoyment of the workplace. In time, as you make your journey from first tentative days to mature role competence, your leadership muscles will need to be developed again and again. So don't skip ahead, thinking my stories of captaincy are for someone else or you at another time. Read, reflect and learn *now*.

Having laid bare how I did not keep myself in best order, I leaned in to learn more and after reading many books, listening to many podcasts and observing many people I found my mantra for captaincy and for my own leadership in general. It was a simple trinity: communicate, clarify and motivate.

Aligned with this, I found my own personal goal, which was specifically made clear to me through Adam Grant's fabulous book *Think Again: The Power of Knowing What You Don't Know*. Grant introduced the term 'confident humility' as a better way to walk through life, and this made so much sense to me. Before learning the term, I didn't have the words to guide my actions. I did not want to be arrogant, but nor did I want to go through life with overt modesty or humility, apologising for any achievements. The notion that I can hold a strong conviction and have confidence in achieving an outcome and in parallel maintain an awareness that I will make errors and need the support of others was perfect for my captaincy. I would hope that I was beginning to behave in this manner regardless, but words matter, and this book gave me the term I needed.

Prior to gaining these insights, I talked in long form about my role and that of the captain I wanted to be, almost in a third-person narrative that was removed from the daily actions that would deliver on this promise. Having read Grant's book, I understood that I needed to ditch the lofty words and build a practice that meant that I, in the first person, could deliver what my teams needed. So, having identified the importance of the trinity, I now needed to understand the actions required to execute each one.

To say 'communication is king' is no great revelation, but what does it actually mean? When we speak of a 'great communicator', what are we really referring to? Are they an orator, are they a person of words or are they open and engaging in a private setting? In what will become a common theme, captains do not have the luxury of honing one aspect ahead of or in lieu of others. The obligation is to speak publicly and privately to people spanning a range that includes the world's business elite, world leaders, dock workers and housekeepers, often all in the same afternoon. Written communication is equally diverse: emails need to be sent to elites working in English one minute, and the next to port agents working in their second or third language.

In parallel to the external challenges of landing the right tone at the right time to the right recipient is the critically important task of doing all the above when communicating with your internal stakeholders, the crew. Sometimes answers to great problems hide in plain sight and for me the best guidance came from one of the cornerstones of the modern maritime environment, the International Safety Management (ISM) Code. This outlines a list of the delightfully simple yet powerful set of obligations that are placed on the captain:

1. Implementing the safety and environmental protection policy of the Company
2. *Motivating* the crew in the observation of that policy
3. Issuing *appropriate orders* and instructions in a *clear and simple manner*
4. Verifying that specified requirements are observed
5. Reviewing the SMS and reporting its deficiencies to the shore-based management.

Points 2 and 3 my favourite in their clarity. Captains are legally obliged to inspire, chastise and console effectively and efficiently. Not many industries have such a simple playbook, and whether you are crew, captain or somewhere else completely, consider for a moment how to adapt this set of guiding notes to your own environment. Do you motivate towards the goals of the organisation? Are your communications clear and simple?

The only shortcoming of an otherwise inspired code is that there is no specific training for captains to be inspirational communicators. Courses in navigation, maritime law and stability do not give any hint of how to improve professional communications. I realised my shortcomings and tried to cover by myself what had not been taught in maritime college. I found I had small wins, but I could not maintain all of the strategies to the expected level at the same time. This wasn't some lofty goal – it was the basics of what was required. I had to go deep into my earlier training to gain some guidance. This training is shared in the Toolbox section on pp. 127–231, but before we get there,

I need to move on from documenting the trials of my captaincy and steer us to a great place: your best superyacht life.

> ### KEY TAKEAWAY
>
> - My personal mantra for captaincy and for my own leadership in general is: communicate, clarify and motivate. Ask yourself whether your words and actions motivate others to meet the goals of the organisation, and whether your communication is clear and simple.
> - As a captain, you must communicate with a wide range of different people, some of whom will be speaking a second or third language. Keep your communication concise and clear, and strive to strike the right professional tone. Toolbox 1 covers this topic in more detail.

7

LIVING YOUR BEST LIFE ON A SUPERYACHT

HOW TO MASTER THE WORK–LIFE BALANCE

I glossed over my emotional state in the months after my earlier-than-planned departure from the yacht in the last chapter. To say I was crestfallen is an understatement: I had fallen so far and so fast I didn't know where I had landed. I had been so proud of the yacht, the crew and my position that I had drifted: I had allowed my self-definition to be linked to my employment and I had not kept anything aside to remember who I was without the yacht, its owner and the public recognition of the position. Who was I without the job? Captaincy is a tightly defined and understood role, as so many others are, which can make many of us question who we are without our professional identities. Few of us take as much time to define ourselves outside of our professional environment as we do within. Later, I will be speaking about values versus beliefs, since this loss of definition was the foundation story for this exploration.

Before I discovered the importance of values and beliefs to support self-definition I crafted a narrative as a sticking plaster to patch over this loss of professional self after I left the yacht. It was true enough on the surface and rolled off the tongue, so it seemed to do the job. When asked, my response to what I was doing was, 'I am taking a sabbatical year to be with the family. I gave a lot in the past and this is a great time to cover for some of the time missed.' It was true, and so long as it was not challenged, it was OK as a cover story. My internal voice was far more confusing: I was uncertain of the future and whether I could manage the demands of captaincy, my family and my sense of self in the future.

I knew I needed to change my approach. The intensity I had been operating at was unsustainable for me, for the crew I sought to lead, for the yacht's owners and ultimately for my family – they needed me to be emotionally available. I had not yet met the wise captains from my future who had learned to place boundaries within their work environment. The first thing I thought of was creating a work–life balance. It was spoken of everywhere, and it seemed it was what I should aspire towards. The first task is to define the aphorism. What is a work–life balance? It was presented in every shiny magazine and their electronic equivalents as the goal, usually accompanied by a linen-wearing parent, smiling and playing with their children all while looking dreamily at the camera. It is a close cousin to the 'happiness' industry and equally as damaging.

Deconstructing the social media myths to get a clearer visual, I see a set of traditional scales – in my mind's eye, the old cast-iron beam scales as held by Lady Justice, at arm's length with her sword in the other hand. Any weight on either side will tip these scales and so will need to be removed or countered with another weight on the other side. This results in a constant tension between the opposing forces that is always ready, always waiting, to be thrown out of equilibrium. Somehow, the marketing hype tells us this perilous, fleeting, ready to be disrupted moment is what is relied upon to deliver us greatest productivity, happiness and health.

As seafaring crew, I do not believe there can be such a balance and seeking it builds more pressure as you try to resolve the tension. Going to sea and living in your workplace immediately tips the beam scale and this is before any workplace or 'pleaser pressure' is loaded in. I let go of the balance model and searched for a new approach.

The first step was to challenge myself and ask whether I really wanted to be on board. Having answered yes to this, then within the reason of physical capacity, I would give all my energies to the role. I wouldn't be burdened by the need to take time for me or keeping a balance. My logic would be that the energy I put into my captaincy would flow across other areas of my life. This reads like I was diving back into the spiral I had only just left behind, but I wasn't. I carved out space to maintain myself – time for physical, emotional and mental care. This was space I had not previously prioritised. To achieve these goals, I needed a plan. As often happens, just when I was looking for a frame to hang my ideas on, one was handed to me. Cue the SHED.

How to organise your SHED

Immediate changes were to ring-fence 15 minutes in the morning to stretch and, wherever possible in the afternoon (a smaller fence), build in 20–30 minutes of general exercise. I had a standing request for a pot of ginger tea to be brought to me on the bridge each morning and I asked the galley to portion my meals. Sorry, this is a captain's indulgence and does not work for everyone – maybe don't ask for this in your first weeks – but these requests did not burden others and the crew knew they were a part of my self-care, which in turn made me a better captain for them. Additionally, it gave them a sense of control over their captain that they enjoyed.

I also handed my radio and normal phone to the bridge at night; I allowed them to be the filter for all calls, including ones from the owner and guests. This benefitted multiple people in several ways: I had a filter to the calls, so that only emergencies would be passed to me; and the guests had their phone answered by an awake and attentive deck officer, who normally dealt with their query directly or on the rare occasion they did want me, would rouse me and I would return the guest's call within a minute or two, having had the chance to wake up and gain a short briefing before speaking.

I know that not too many workplaces involve calls through the night, but on a yacht you can get many non-filtered invasive requests and demands. By placing reasonable boundaries around your time and energy, your reports up and down the line know where you and they stand. It might at first sound counterintuitive, but by placing some lines on your personal court, the other players know where to hit the ball. The restrictions will actually make you more available to them.

In introducing this shift in approach to my team, I spoke of the framework that I was using. Around the same time as I was trying to understand what changes I needed to make to not be underweight and strung-out on a remote beach, I was listening to and learning from a wonderful keynote speech by former Royal Marines Major General Andy Salmon. Among many substantial insights, he spoke of the SHED. This is so simple to remember, but a little harder to live.

This is what your SHED involves:

- Sleep – stable in amount and quality
- Hydration – learn your needs; monitor your consumption
- Exercise – as much as you can fit into your day
- Diet – diversity, simplicity and portion control

How to get your emotional life back on track

With my SHED beginning to organise itself I was in a better place to think of the other people in my life. My former self never set a time to speak to my wife Yvonne and I never pushed my daughters Fabienne and Scarlett to tell me about their school days, and consequently often one day became three where we didn't speak and on other occasions three became five. If spontaneity is fun, then routine is the backbone of success.

With advice from Yvonne, I set alarms for times when the girls were most ready to receive a call. Given that I worked across multiple time zones, this took some management, but the outcome was we all began to recognise these scheduled calls and were ready with our stories to share. The process of re-engaging and reconnecting was working. Sharing photos with the family was a task that was not time-dependent and I started to do this with more regularity. My selfies are terrible but are better than nothing and in return I received more updates from home.

In addition to my family, the other intimate relationship I needed to maintain was with the crew. Onboard work- and task-related communications still filled the days, but in the evenings, I would drift to the bridge, engine room or crew mess and just sit and meander through a conversation, using it as a chance to ask questions of others and learn about their lives a little more. There were some stellar experiences among these evening chats. The circle was beginning to reinforce itself, my flywheel increased in pace, and my outcomes were improved physical health and social connections. With the fundamentals in place, I had the capacity to move to the next level of personal sustenance and development.

Weightlifting for the mind

Thirty minutes of reading each day aligns me; it allows me to walk along an avenue of words laid out by an author, to stop and look in the windows of the places and people they describe. This experience is not available automatically; curating books is important, and I enjoy this too. As I near the end of a book a nervous anticipation builds, not only for the culmination of the story, but for the excitement of choosing the next book.

All leaders and all members of a successful team must lead, motivate and inspire consistently, whether they feel like it or not. The ISM code's delightfully simple obligations reinforce this in the maritime sector, but with or without this, the concept is universal, and it takes an energy greater than you feel is available to you and more knowledge than your own single journey has provided. I do not hide my reading habit: I share openly that the books I read inform and guide me. They give me new perspectives, they stop me being stuck in the tracks of my own cognitive bias and they underwrite or challenge my assumptions. Of course, I am not speaking of a particular book or even books that deliver; I am talking of embedding the behaviour of reading to learn in your daily leadership practice. I will go further and state it more boldly: you will not be the leader you want to be and what those you lead deserve without this consistent learning.

Habits are repeatable behaviours, and there are few more rewarding habits than embedding reading into your life. Start gently, and with diversity. Read a crime thriller in between some non-fiction, read quality historical fiction to inform and entertain at the same time. My own process is to use an electronic reader that has a highlight function. I highlight text for both future reference (a fact I wish to recall) and for the beauty of the words (just for enjoyment). I then look through these highlights and rewrite them by hand in my journal. I really like this process of writing out the highlights – not for the leveraged learning it provides, but because I have two opportunities to enjoy the words. Some days it will entail 30 minutes of writing; on others it will be only five minutes if my mind is distracted. There is no pass or fail; I win just by engaging in the process. As Cicero says: 'Read at every wait; read at all hours; read within leisure; read in times of labour; read as one goes in; read as one goes out. The task of the educated mind is simply put: read to lead.'

'Read at every wait; read at all hours; read within leisure; read in times of labour; read as one goes in; read as one goes out. The task of the educated mind is simply put: read to lead.'

Process over performance

Through the momentum of my daily process, I am now more firmly placed to maintain the intensity that captaincy demands and be protected from sleepwalking into burnout mode.

In addition to maintaining the daily repetitive processes of the SHED I have found one other special, and until now, very private practice. I look for a thing of beauty in each day; it is everywhere, if you look. Some days it is overt – a glorious sunrise or sunset – and other days it requires some deeper investigation, but it is always there. Yachts provide many avenues to find beauty, and when days are long it is worth highlighting these to other crew as the privilege of our employment. Before the non-yacht reader derisively *hmmph*s at my finding beauty from a luxury yacht at sea, I say that more authentic daily beauty can be found everywhere, from the train to the supermarket – you just must look. Doing it is easy, and it can help you to be present in the moment and provide stress relief.

The processes I began to lock into my daily routine were not chosen at random: I experimented, and found that some things took too long or were too complex. They had to be repeatable, enjoyable and deliver the energy I needed to draw upon when the days felt long and hard. Somewhere around this time I stumbled across a quote from the sculptor Elizabeth King that seemed to sum up why I was doing all this, 'Process saves us from the poverty of our intentions.'

'Process saves us from the poverty of our intentions.'

My intention was to be the best captain, the best leader, the best crew member I could be, and the process provided me with the platform to realise this. I stopped looking at the tasks of each day as work: I looked at them as experiments that I was conducting in my own life laboratory, each experiment aimed at increasing my performance. The humble email became an experiment in better communications. Could I improve the presentation? Could I add some value in the recipient's day? A conversation with the crew became a chance to learn more about them and their workplace. What could I learn in the laundry today? What can the watchkeeping engineer teach me? What questions work well to gain a flowing response from an otherwise shy, reserved crew member? My aim was to get three or four questions deep in every conversation, to get behind the banal surface questions to deeper communication.

No, I did not achieve this level of positive curiosity every day, but knowing it was part of my process helped me realise what I was aiming for. And like a navigation track on a chart, I knew when I had to correct my course to return to what I had planned. Stretching the metaphor, I could also recognise when I was off-track to the point of hitting a predefined cross-track error. This might look like three days in row with interrupted sleep, or two days in a row during which my enabling processes were cast aside. With these metrics in my consciousness, I knew when I was heading away from being the best I could be and what I needed to do to course correct.

The value of this was that when I returned home, the momentum I had built with my onboard processes flowed into that space too. My family no longer welcomed a tired shadow of the husband and father who had left some months earlier. What walked through the door was someone full of stories about the days, the sights, the other crew they had worked with. The cumulative effect of these small shifts delivered me more energy than it took to establish the behaviours. It almost felt like I was cheating. It was sometime later that I heard the author Jim Collins speak of the 'flywheel effect', where the momentum of your efforts delivers more returns than the energy you put in to sustain it. To this day, I believe this works for me.

INSIGHT

Best crew life – success on a superyacht – is not about balance: it is about giving it all and using the energy from this to feed into your life.

If you don't love the work, don't start the journey.

KEY TAKEAWAYS

- Organise your SHED. You need to plan your sleep, your hydration, your exercise and your food, especially when you are on board a yacht. Otherwise, they won't happen. If you have less control than I do over these elements, it is still worth taking control over the parts you can.

- Maintain and nurture links with home. Plan for when and how often you are going to contact your family or closest support person and stick to it.
- Aim to read every day, across a diverse range of genres and topics. It provides a mental escape and will greatly broaden your mind.
- Finding or noticing 'glimmers' – be it a thing of beauty, the taste of a good cup of coffee or the sensation of the sun on your skin – helps you to be more present and to maintain a more positive mindset.
- Working out which processes work for you on board and in life in general might take a bit of trial and error. To become habits, they need to be repeatable, enjoyable and deliver more energy than is required to do them. You might lapse some days – we are all human, and things happen than can derail the best-laid plans – but once you've realised the benefits, you'll soon course correct and resume your good habits.

A year in the wilderness

I have said above how a few easy shifts created a significant and lasting change in my life to the point where everything was shinier, a better version of its former self. If only it were that easy.

My words of taking a break, learning a little, jumping back in, turning a flywheel, and making captaincy work on my terms I hope reads well, but are they my truth? In hindsight, there is more truth than faded memory in there, but during the living of these words there were scary times. I suffer terribly from male provider anxiety, a sense that I need to look after my family financially, whatever the personal cost. This is not a sexist term: I do believe both sexes feel this, but I can only experience it as myself, a male, so that's the only perspective I can give.

This manifested in my doubting I would ever gain employment again and darker thoughts moving further in this direction. I knew the feeling: it wasn't the first time I'd experienced it. Earlier in my career, Yvonne and I were working together. We took the pending sale of the yacht we were both employed within as a sign to move on, so we planned a life-affirming trip to Northern India where Yvonne's brother was working. I didn't make it to India with Yvonne and her brother. The phone rang, there was a yacht needing a first officer and due to my insecurity about always needing employment, I took the position. I still regret not sharing the epic journey through the northern reaches of an amazing land with Yvonne. I was a product of a conservative upbringing where employment and providing for a family were the cornerstones of a stable community. This was never assuaged through time in the military or working in the commercial maritime sector; there are not too many freewheeling types drawn to these organisations.

I would be amazed and somewhat envious when surrounded by crew mates who moved lightly through their world with seemingly little concern for their longer-term security. I wanted to emulate this: I wanted to just step off a yacht and drift around Europe or Asia or ride a bike across the USA, taking each day as it presented. These were delightful dreams, but my character caused me to continue my education and career in lieu of more carefree options. I learned to not covet a life that was not in my nature, and I moved ahead in my career with planning and thought.

I recognised captaincy as my goal, and I knew what to do to get there. When others partied, I studied. When exposed to enablers in the industry, I made an effort so that they may recognise my potential value to them. As I have detailed, a combination of hard work and some good luck in timing saw me move swiftly through to a position where I could start looking towards my first yachting captaincy. I had benefitted from some exceptional role models, but I had yet to find my own theme, my own voice, my own guiding philosophy. I was still not ready to fully support others, as I still needed to develop myself; I was only just beginning to pull the oxygen down into my own mask.

Internal struggles of self-worth aside, the time after I separated from the yacht in 2013 and was unemployed was not all terrible.

The text below was written on completion of the 'sabbatical' and in the years since I have looked back to reflect and remember. I called the document 'a year in the wilderness'. This made it difficult to find again on my computer, as it didn't really refer to anything, but I never changed the name because it spoke to how I felt at the time. Below is the text I wrote for myself:

Twelve months ago I was a loving, though largely absent husband, and father. Even when physically present I was emotionally elsewhere. Through events not entirely under my control I finished work on 18 February 2013. A friend (thanks, Michael) asked me not long after, 'So how long will you be taking from work?' Having not really thought about it, I shrugged and fumbled a response. He continued, 'You should set yourself a not-work-before date so you will relax and enjoy the time.'

This made sense and I went through some dates, and nothing leapt out. Within days my then seven-year-old daughter said, 'Papa, I know it wasn't your fault that you missed my birthday this year, but could you stay home until my next birthday?' Her birthday was 3 February – perfect! I'd found an answer to my timeline dilemma and having missed almost all the family's birthdays for many years due to work commitments, it gave the chance to have a clean sweep in one calendar year.

With a year free of work, I thought I would:

- *Cycle tour somewhere extravagant for a few weeks*
- *Read*
- *Really improve my German*
- *Complete writing the book that a publisher had shown an interest in*
- *Read*
- *Kayak the Donau (Danube)*
- *Single-handedly complete the development of a property we had bought*

Now with the year past and clear I didn't cycle or paddle anywhere of note. I did not get past chapter outlines for the book, and the property will be started by the builders this coming February. My German is better but not great and I read a bit more than usual but not so much as to be impressive.

Was the year a failure? I have read a few articles about 'bucket lists' and families doing amazing things during their 'sabbatical year'. These made me feel self-conscious about the little I had done. To assuage this feeling of underachievement I made a list of my thoughts and conclusions. I have shared it below to enable me to remember the year and for others to realise that you too can achieve amazing things – all you need is time.

My Year

1. *Good conversation requires more listening than talking.*
2. *Trying to be the smartest guy in the room will result in a lot of time standing in empty rooms.*
3. *Triple sifting flour does make a difference when making muffins.*
4. *I like parody music clips on YouTube.*
5. *Spotify is great, but too much choice is challenging.*
6. *My eight-year-old daughter is a better guitar player now than I will ever be.*
7. *Asking personal questions of friends and family is not rude; it shows you care and is appreciated.*
8. *Follow-up questions are good too; most people answer your first question hoping you will ask more.*
9. *Making interesting breakfasts for the family is rewarding.*
10. *Social media is OK – so long as you are its master and not its slave.*
11. *Checking mobile devices during meals and in conversations is unfathomably rude.*
12. *Helping family and friends at their houses is hugely rewarding, even when the work is grim.*

13. *Not working is a great test for who liked you for you in the workplace and who liked you because of your job. Embrace the former; discard the latter.*

14. *Climbing mountains or proving your worth in other solo pursuits means nothing to your children, family and close friends. Time together means everything.*

15. *Starting new businesses is tough. There are no 'free rides' out there and most markets are quite efficient (see comment on smartest guy in the room).*

16. *I like John Denver.*

17. *I don't understand what # in front of a word means.*

18. *My wife works hard for our family in our home. She just doesn't go on about it like I did about my former work (my work was not that hard).*

19. *My friends are great. They have cool families; I just needed the time to meet them.*

20. *Fixing things is more rewarding than replacing them.*

21. *Romcoms are OK when watched with your partner or children (so long as your manfriends do not find out).*

22. *I like being the house guest of my friends and I like having my friends as house guests.*

23. *A week as a solo father with children is worth six weeks of normal parenting (approximate ratio).*

24. *Spending an afternoon making a meal for others is a great use of time.*

25. *I like the Voice TV franchise.*

26. *Receiving two emails from friends is equivalent to 100 work emails (approximate ratio).*

27. *People who can only talk about their work bore me.*

28. *Judging yourself by your employer's happiness is a road to discontent.*

29. *Children are complex; they're experiencing the world so fast that they need attention on the journey. But it must be their journey.*

30. *My nephews are grown men and are different to me, and that is cool.*

31. *Giving attention to children and maintaining a house is a full-time job. If you can downsize your life to allow one parent to do this, you will be richer.*

32. *You need that new car much less than the family needs your time.*

33. *I walk slower now than a year ago.*

34. *A year is too short to stop work and the big goals that we are told we must follow are sometimes less important than diligently applying yourself to the small daily challenges of life.*

35. *Having the time to let another motorist into a traffic queue or a person into the grocery line is so easy and yet so rewarding.*

36. *Going back to work at the end is not so bad.*

Reading this list a decade later reminds me of that period. I see some points that reflect my state of mind during the time and others that seem more timeless. More important than the specific words on the page was that I wrote a record from me, to me. I find journalling awkward, as I run out of things to say, but making a record of accountability for this stage of my life was hugely valuable and is one I recommend.

FINAL WORDS ON LEADERSHIP

I have walked briskly through the stages of crew superyacht life, as this book is not written to provide all the answers. Indeed, it could not if it tried, since the answers are yours and will change with time and circumstance. I do hope my words may make you ask better questions and challenge the answers with increased rigour.

I have not hidden some of the pitfalls of a career in the superyacht community, especially when you are a leader. I have intentionally leaned on a few of them to gird against the giddy emotions that the beautiful large yachts evoke – in crew, their owners and most who see them. But the life can be fulfilling, rewarding and just that little different from the mainstream. Your superyacht life may not extend to decades; I thrive on crew who openly say they are only going to work in yachting for a few years for some fun and travel and then they will return to their 'normal' lives ashore. Bravo. If this could be the long-term career for you or the family member you are reading this book for, though, then please do take note of the risks, the boundaries and the cautionary tales I have shared.

PART FOUR

THE BUFF PAGES

The *Reeds Nautical Almanac* is the inspiration for this section. A book that has been in continuous publication since 1767, the *Almanac* is now more used by recreational sailors enjoying their connection to the past than superyacht crew, but once it was critical for all mariners. It contains a wealth of information, and has all the numerical tables to turn split mirror observations of celestial bodies taken by a sextant into lines of position on a chart. This was a fabulous tool before computed solutions and the advent of the global positioning system. Within the book there was an even denser part called the 'buff pages', which gave me the name for this section. These pages were printed differently and contained the daily essential tables and astronomical conversion information. I hope my toolbox pages can help you in some way like the *Almanac*'s buff pages helped me in my junior days learning to navigate by the stars.

Toolbox 1 – communications

CRYPTIC CHALLENGE – THE IMPORTANCE OF GOOD COMMUNICATION SKILLS

Throughout the book I have spoken of the crossover between all aspects of my career. I would like to think I do not wear my military training on my sleeve, but I do draw upon it, and especially so when it relates to communications. Officers in defence forces are not taught how to 'do things': they are taught how to lead others to do things. Their leadership acts as a force multiplier, and applies as much to superyachts as it does to the military. To illustrate its application in both arenas, I'm going to tell you about a training experience I underwent as a cadet.

All cadets at the Australian Defence Force Academy in their second year of training take part in an event called 'Cryptic Challenge'. It is a major assessment, and we were made very aware that our performance in the exercise would impact our future within the Academy – performance in the Academy being directly correlated to longer-term military career opportunities. It is also a logistically heavy event on a scale only the military or reality TV would invest in. In my year, three different challenge weeks were run consecutively with approximately 80 cadets per week. In support of these numbers of cadets was a matching staff: observers, medical teams, caterers and logistics types to ensure the trucks and all the paraphernalia was where it needed to be, when it needed to. In the months leading up to the challenge, we received specialist training from medics, mechanics, combat instructors and radio technicians, who focused on the technical skills that would be needed to overcome the challenges and, more importantly, leadership and teamwork strategies so that we would work together. We were also prepared in terms of fitness, with increased levels of physical training.

After all the anticipation, very early one cold Canberra morning we lined up in our teams beside a fleet of military trucks and buses. I remember well trying to maintain a swagger as we placed our packs in a gear truck and then continued to the transfer bus. I sharpened up when one of the sergeants crisply cut me short with, 'You won't be joking around when you haven't eaten or slept for 48 hours.' My head dipped and I boarded sullenly. We had heard rumours about the event, but it changed each year, so one group's experience was not completely transferrable to the next. Over a seven-to-nine-day period (the exact length was kept from us) the teams of eight would be required to undertake a series of tasks. All instruction and equipment would be available in some manner, but the organisation of humans and equipment would be the point of success or failure. Roles within the group would rotate so that each team member had the chance to lead, observe and, of course, participate.

The teams of eight would be 'on their own', the premise being that the situations we were asked to resolve happened in a remote area without external support. It was intentional that we would feel isolated, physically and emotionally. There was always a staff observer noting performance; at times they would be in sight, but more often they would not. As our fatigue built we would lose awareness that we were in a managed environment. If the goals were to make us feel alone, fearful and inadequate, then from my perspective the event overachieved.

There are a couple of important details that need to be explained. The challenge varied depending on the team's performance. It was not about the technical execution so much as the communication that led to it: if we did not follow the process, the directing staff (DS) would intervene. The solution to some of the tasks was quite evident and if the team automatically began 'solving' without formulating a briefing and a leader-led plan, then a new piece of complexity would be radioed in to make it more challenging. In parallel to the specific tasks was an underlying requirement to navigate and orienteer through the forest where the challenge was located, and to do so at pace. Checkpoints needed to be met through time management or, again, the DS would change the rules.

Eighteen hours into the first day, I had not eaten, and sleep was not in sight. My legs were tired, and moving was getting harder. We had a pace to maintain to reach the midnight food drop, but we were nearing the stage of having to make the decision to slow down, conserve energy and acknowledge we would miss the meal. We discussed sending ahead a 'runner' with no pack or rifle to reach the food drop, but this was discounted. The number one rule was to maintain the team's composition as a unit. We assumed there would be eight challenges, to match the eight cadets in the group, but the staff had warned us, 'There may be more; there won't be less. The total depends on your performance as a team.' We had heard the stories from those that had gone before of additional tasks and additional days, so we were not taking the comments from the directing staff lightly.

We had been walking, navigating towards a reference point, when our path intersected with an abandoned Land Rover in a creek bed. In the cabin was a briefing note informing us that the vehicle needed to be moved across the creek and up the embankment to the forest road. A recovery vehicle would be dispatched once we radioed in upon completion. As we inspected the vehicle it was clear that all the equipment we needed to move the vehicle was there: ropes, chain-blocks and snatch straps. What was not included was information on how to rig them, and to ensure the task was not too simple, there was much more equipment than we needed in the tray of the vehicle. Some debate ensued within our team as to how to set the ropes and blocks. The team leader, with no sense of rigging blocks, was trying desperately to use knowledge only gained from the classroom. With some prompting and some circumvention of the leader's authority, we moved the vehicle and continued.

After another three hours of walking, we had discovered a screaming 'injured pilot' lying in a field. The pilot was non-communicative but, again, there was a briefing sheet. He had been ejected from his plane and had to be moved to a new point, where a medical team could evacuate him. He was injured, with a broken leg that needed splinting. This was not so challenging for the assigned leader. We all knew how to make a leg splint and had been shown how to construct a stretcher from available materials – in this case the parachute conveniently remained in the area. We moved our pilot in quite good time and were again following instructions for the next challenge.

This setting of challenges that relied on leadership and communication for their successful outcomes continued for each of the days, until we completed the challenge. I would like to say I enjoyed the event during its execution, but I didn't. I was tired, confused at times, and frustrated by the process of leading and being led. I was trying to draw upon classroom learning while not losing a commonsense approach to solving the problems. We missed a couple of meal drops due to the slower members of our team and I found it hard not to resent their impact on our (my) performance.

I think about the Cryptic Challenge often – how well conceptualised it was and how well executed. I wish, costs aside, I could do it again and test my growth or regression over the years. It would not be the physical tests that I would seek to check competence with; successfully pulling the Land Rover from a ditch was never the point. I would assess my ability to deliver a clear, structured briefing, without the benefit of the focused training and the structured environment that the challenge and the military framework provided. Communication is the absolute foundation in the captain or leader's quiver of competence. I know at times I do OK, but there is always an awareness, a self-consciousness, that I am never replicating the standard I was lifted to during that Cryptic Challenge. The ISM code referred to earlier (see p. 107) had not been written when I was undertaking the Cryptic Challenge, though the expectations of the cadet team leader role mirrored that of the code's obligations to captains: identify and uphold the policies (plan); motivate your team to follow them; and issue clear and simple instructions.

Key skills

The military are not alone in their delight in using acronyms, but there is one that I have continued to use in my life after completing the Cryptic Challenge: SMEAC. I have lost count of the times when I have been confronted by a complex situation and SMEAC has helped me break it down to achieve my goals. Let's take a look at what it stands for:

Situation
This is stating your problem, or a description of what has occurred, to see you where you are. What were the events leading up to where you are now?

Mission
Once you know what has happened or have defined the problem, the mission is stating what is to be done. It is a concise statement of action to achieve the solution.

Execution
This is the breakdown of what you are going to do, and should involve the subheadings: What? Why? When? Where? Who? How?

Administration
This is the resource allocation section. What do you need, in terms of time, people and things?

Command and communications
To successfully complete a group task there needs to be a commander – a decision-maker or, if the project is big enough, a hierarchy that defines reporting and accountability. This is where the communications are defined – something that has become evermore important as the choice of software used to transmit that communication has become as essential as the project itself.

> ### Key takeaways
>
> - Clear communication and co-operative teamwork are vital to the successful completion of tasks, especially in challenging situations. There also needs to be a clear leader so that everyone knows their role.
> - When assessing a new situation or challenge, try breaking it down using SMEAC. This will give you clear oversight and everyone will know exactly what they are supposed to be doing, and to what end.

THE WEDDING SPEECH – HOW TO COMMUNICATE EFFECTIVELY

There are countless guides to effective public speaking out there, so I will keep my own advice brief. That said, I do suggest that you watch some clips of great speakers and look for the cues that match your own style. In the modern era it would be fair to say Barack Obama represents a holy oratory grail, but that was his voice and his style, it would be foolish to try and copy this. All speakers need to find their own voice, one that is authentic to them whether in a group of five or five hundred. In my first book I spoke of visualisation as key to successful performance during a fast-paced emergency, and this also applies to public speaking. Poor performance will be a self-triggered emergency, so alongside researching great speakers, there are baselines that must be followed to avoid failure. I'll touch on these now.

I am not a natural when it comes to speaking in front of any group, so before any occasion, from a crew mess to an auditorium stage, I will visualise how I will be doing it. This includes where I will be, how I will stand, where my notes will rest, and whether I can see everyone in the room in one eye sweep. If it is a crew talk, I think about whether the tone will be warm and comforting, whether I will be seated and acting as 'one of you', or else being firm and standing up to add gravitas to my words.

I will also practise my talk many times to myself before I perform it in front of others, and ask myself a few simple questions. Where are the pauses? When do I weight a word or repeat it for effect? The first time I stood before a mirror to practise a speech I felt ridiculous, and the same was true when I began recording talks in advance and watching them back. Yet I do it because I know I need to. This rehearsal is essential for me and without it my performance suffers.

Just before I give my talk, I need to regulate my breathing. I take three to four excessively deep breaths (out of sight of others) to steady myself and allay my feeling of anxiety. I also do some warm-up exercise before I enter the room. I have a favourite YouTube tutorial for this, but there are many to choose from and I recommend finding one that you can watch without cringing too much. The one I listen to opens with, 'It's probably best to do this alone' and I agree. The contortions and the tongue twisters work well but not if viewed by a third party. This may suggest to you that I am unusually anxious about speaking. I am not; these practices are guided by me seeking to perform at the highest level – not the Obama level, just my highest level. I know from experience that they are well worth the investment of a little time and preparation.

Key skills

Too many speakers think they will be fine just getting in front and having a go. It if were this simple, there would not be the hundreds of books, talks and tutorials produced on public speaking and nor would many people rank it as their greatest fear. In my experience, yacht captains are less inhibited by this fear than the wider populace, perhaps because they're emboldened by the stature that is perceived as being part of their role. Sadly, this does not always translate into sparkling oratory performance – it's sometimes more a rambling noise, like the strange uncle at the wedding whose speech leaves the room awkwardly squirming in their seats with their shared embarrassment. Do not be this person: be the captain who comes in with purpose, speaks clearly using a prepared set of notes, and leaves the room informed and uplifted.

The second point here is critical: the content is important, but the delivery must override this. The crew must gain energy from your presentation. You are giving inspiration, not taking it. Bear this in mind when you consider your content. Too many times, I spoke in crew meetings on mundane topics that would have been better shared on the crew noticeboard or in the ubiquitous message groups. Instead, use public speaking as the forum for hitting the high notes, to drive home to the crew that you are there to be part of the greatest team ever to sail. If you are not the greatest team ever to sail on a superyacht, then inspire them to work towards that goal.

Avoid the bully pulpit. The only thing worse than the 'strange uncle at the wedding' captain or person in a position of authority is the one who uses their unopposed right to speak as a chance to berate and belittle. There are times when the yacht needs a firm course correction, and it is the captain who should do this, but remember, if the yacht is off its cultural course for whatever reason, it is by your hand. Speak firmly, read the landscape truthfully, but with you on the ground with the crew, not loftily soaring above talking down. As the captain, you can control all of this. You can prepare, visualise, practise, watch others, develop an authentic voice and even deliver chastisement with empathy.

What you cannot fully prepare for is questions, so you just need to try to anticipate in advance what may come up and think through some answers. I can recall so many times in captaincy when I gave what I considered a professional presentation to a group and then undid it by handling the questions poorly. Regardless of the audience size it takes courage for someone to ask a question, and that courage needs to be respected and rewarded. A consistent failing of mine was to race towards what I perceived as the answer, virtually cutting the questioner off in the process. In my own case, this was born from a need to retain control. A better way is to think of the person asking the question as though they are adding value, not taking control from you.

They have listened and now seek to challenge or share something that is of importance to them or to their colleagues. Their motives may be open to question, but that is not yours to judge in the moment. I must work hard to overcome my intuitive response to 'quieten the speaker, answer the question and move on'. When I do tame my instinct, it always goes well. If I pause to listen to the speaker, repeat the question and respond with sensitivity my overall performance improves. Importantly, I remove the pressure to even answer the question. More often it is entirely acceptable to acknowledge the validity and commit to answering later, whether that is in private or the next public setting.

INSIGHT

Captains, like all leaders, must be able to deliver a clear message to their team: getting it right will lift, motivate and guide; getting it wrong will separate, flatten and confuse.

KEY TAKEAWAYS

- Come into a meeting or room with purpose, speak clearly using a prepared set of notes, and aim to leave the room informed and uplifted.
- Consider the content of what you are saying. Could it be better placed on a noticeboard? Save meetings for big-hitting topics and for motivating, rather than transmitting minutiae.
- Do not use a position of power to bully people and deliver a tirade of negativity. Seek to deliver any chastisement with empathy.
- When taking questions, show care and respect to the person asking. They are more nervous than you are. It pays to think through in advance what some of the questions may be, so you have some prepared answers, but always expect the unexpected.

HOW TO GIVE EFFECTIVE BRIEFINGS

As a developing captain I thought I was being judged by the quality of my ship-handling. I thought when I squeezed what was becoming ever-larger yachts into seemingly ever-shrinking ports that this show of professional skill was earning the respect of my crew. Some might pass comment, but fundamentally, ship-handling was an expectation for the captain's role with no kudos associated to it. Like an engineer servicing an engine, a chef making a meal or a housekeeper cleaning a cabin, ship-handling is not where a captain is judged.

Where crew did pass firm judgement was in my ability to speak, provide content and inspire them when they needed it. Above, I talked about the importance of preparation for public speaking, and crew meetings are foremost in the captain's armoury of communications.

Key skills

There is no single how-to when it comes to giving briefings; much is shaped by the geography of the room and the topic. Sometimes it is appropriate to stand, schoolmaster style, and equally there are times when sitting among the crew and speaking gently on a sensitive topic are what is needed.

Try to avoid reading throughout a briefing, but do use notes to provide structure. Remember, if you have called a group together then you must show value for the time taken out of their day, which is multiplied by the number of people in the room. A one-hour presentation to 40 people equals a normal working week for one person ashore.

Beginning, middle, end. That's the structure you should adhere to. Begin by introducing what you are going to say. The middle is your message in full and the end is the summary of the points you have made. Depth and nuance can be layered on this structure, but its universal simplicity is what makes it repeatable and transferrable across every topic.

Repetition is another useful tool. The best speakers weave repetition in so naturally that it is present, powerful but imperceptible. They do this under the banner of classic rhetorical delivery, which is a centuries-old structure for delivering meaningful content. It is an area of study within itself, though if you wanted to use some classic rhetorical devices in your own speeches then you should ensure you are adhering by these pillars:

- *Logos* – the argument appeals to logic and reason
- *Ethos* – you are credible to deliver the message
- *Pathos* – you establish an emotional connection with the audience.

Clarity and clarification are key to any successful communication, especially in briefings. The only way to do this is with crisp and concise messaging that is appropriate to the room. Returning to the opening of this book, the mission is to keep the signal-to-noise ratio high.

There are several topics that impact the crew and they recur frequently. These include life on board, the yacht's programme, crew employment conditions and their performance as a team. Think about each one ahead of any group communications and seek insight as to where any confusion might lie. In the case of the yacht's programme, you might not know all the answers, and this is OK. By saying 'I don't know' there is no confusion, especially if you back this up by saying, 'Regardless of destination, I do know we need to keep working together and our mission does not change.' If the message you have to deliver is negative, own it and tell it straight, with no frills that might send out mixed messages.

INSIGHT

There is no word count on good communications. The metric is the ratio of messages passed by the sender to messages understood by the receiver.

HOW TO COMMUNICATE IN WRITING

If the crew mess is where the crew build respect for their captain, then it is in the written word where your shore colleagues will make their judgement. Some years ago I had a shore manager whose contribution to my annual appraisal included the line, 'Writes well, but does so more to impress than express.' I mocked him to myself and saw this as a point of praise. It wasn't: I was communicating under the command of my ego and emotions and he had called me out.

Key skills

I grew, I started to think of the perspective of the reader and how my writing could make it easier for them. What could I do to that would make their day a little better? How few words could I use and how could I present them most clearly? I tried to write without an ego, imparting as many facts as possible without my editorial voice seeping through. The facts are the point of the email. If they want my view, they will ask for it. I made a mental list of dos and don'ts. I would rename emails to assist all readers. I would use bullet points and headers to ensure information was clear. I would not copy in others for the only purpose of showing off or write more than was needed, nor would I use verbosity when a simple message would suffice.

I am still guilty of embedding Latin phrases, references to obscure academics and song lyrics from Australian bands in the 1980s within professional communications, I cannot completely hide my inner tosser.

Key takeaways

- Clarity is key. For many types of written communication, facts are what matter, not your opinion.
- Structure your written communication using clear headings, bullet points and an appropriate format and mode of address.

LISTENING – TO LEARN, TO GROW

Leadership involves many private conversations. These are critical and if there was to be an area of my own career where I have consistently come up short it would be the quality of one-on-one conversations. What was frustrating was I could feel the shortcoming in the moment. I worked upon improving it but never quite achieved the level I hoped for. I picked the low-hanging fruit, I made the setting more conducive to open communications, I knew to ensure digital distractions were removed, but I missed the biggest one.

I had been a captain for so long that it had damaged my conversation. I am good at quickly absorbing information and finding solutions. This serves me well in the fast flow of daily yacht operations but lets me down when it comes to supporting others and working on solutions to longer-term problems. I realised that within the gunwales, it felt easier for me to speak, and others to follow, rather than to take the time to listen to what a crew member was trying to say or contribute, even though this limited the solutions available.

Crew rarely come to the captain to solve their problem. Granted, this may be an outcome further down the path, but crew primarily come to be listened to. Over many years and countless conversations, I listened the minimum I felt I needed to, then turned the person's words around to 'solve' their problem. Thinking back, I was usually pleased with my efforts, and I was often confronted that the crew member did not seem as appreciative of my solution as I was. I had completely missed the point; I was listening only for the cues I needed to 'solve the problem' they had brought to me. What I had not realised is the crew member had a full story to share, and I had cut them off. There is a very real chance the bigger message that they hoped to share was not in their opening dialogue and that my reaction gave them no chance to reach where they were hoping to get to. To address this, I searched for information on better listening. It is a topic fraught with poor information, but within the coal heaps of the internet, there were some diamonds, including this quote by William Ury, founder of the Harvard Negotiation Project: 'We assume a winning negotiation strategy is about talking, when in fact it is about listening.'

'We assume a winning negotiation strategy is about talking, when in fact it is about listening.'

Having discovered William Ury's TED talk, I was then led to Sheila Heen. Heen is a senior lecturer on law at the Harvard Law School and a member of the Harvard Negotiation Project. She is renowned for her work on communications and negotiations. Specifically, difficult conversations – the type that seemed to match about 80 per cent of my interactions as a captain. I listened to her being interviewed before further researching her work. She is an engaging speaker and, while all her points were of interest, there were a couple of phrases that caught my attention: '[We use] listening as a strategy of last resort; we really just seek to give our opinion.' And: 'I will pretend I am listening while I am trying to figure out what I am going to say.'

Ouch! These seemed personal to me, and, for the slightest moment, I thought Professor Heen was speaking directly at me. Her talk identified a couple of my known weaknesses and motivated me to continue in my research. Next, came Jennifer Garvey Berger, who echoed Professor Heen's words, but then defined and framed them further. Garvey Berger spoke of levels of listening that we all unknowingly move through. The first I recognised, the second I understood, and the third was new to me. Without trying to duplicate the tone and eloquence of Heen or Garvey Berger, the levels in order of their precedence were: listening to compete, listening to solve and listening to learn. These are defined as:

'[We use] listening as a strategy of last resort; we really just seek to give our opinion.'

and

'I will pretend I am listening while I am trying to figure out what I am going to say.'

- Listening to compete – responding to the speaker with your 'better' story
- Listening to solve – gaining as little information as needed to 'fix' the speaker's concern
- Listening to learn – a lovely place – taking the time to ask open questions to hear the full story.

INSIGHT

I had spent all that time running around 'solving' crew problems when what I should have been doing was listening intently to pick up all the messages being transmitted by the crew member. More than this, I should have been on the alert to spot guarding tactics on certain topics and ready to ask the probing, open questions.

Key skills

The shift to this way of listening is very difficult and I am still a long way from where I want to be. When something is not natural for me, I will seek a tool to assist. Asking a crew member to, in advance of our conversation, write down three to four points they wish to speak about helps. I will do the same, thinking of points I would like to speak to them about or I that think they would like to raise with me. After the initial settling in we swap our pages. In doing so we have committed to each other that we will explore and hold each other accountable for what we had hoped to achieve prior to entering the conversation. This binds us; we can't back away from something and rush to a conclusion. I will give a minimum of one hour for any of these conversations and ideally keep a buffer after for them to continue. Your day might be seemingly busy, but this is the most valuable use of your time.

In my notebook, which is always with me during conversations, I write 'Listen, reflect, repeat before response' at the top left of the page. In the front of the notebook is a page containing a series of questions that I can flip to without disruption to the conversation. These can be inserted into most conversations and deliver great outcomes. These are a few examples:

'What was that like/how did that feel?'

'Tell me/Explain to me/Describe to me ...' (T_E_D)

'Can you expand to help me understand a little better?'

'What other items are left to discuss?'

'What are the next steps?'

My last communication tool was taught to me by a talented executive assistant to a yacht owner. As we finished each conversation he would, by default, conclude with 'Is there any more I can do for you?' and 'Are there any barriers left that I could help remove?' This method was so impactful that I asked him about it. He said it was the way his former boss (a US member of congress) finished every conversation. He found it so successful that he embraced it as well. I may not be quite so robotic with my own adoption, but I do think it is empowering.

That listening is a key component of better conversations and communication is a given. Knowing when to speak and when and how to listen is the challenge. I have found in my

142

own journey that as I strive to change, I am working against learned behaviours and regularly need to turn to my notes, references and mentors to improve. Maybe your conversational performance is way ahead of my own, but perhaps there is room for a little more improvement?

KEY TAKEAWAYS

- Listening during private meetings is more important than transmitting. By rushing to 'solve' an issue without hearing the other person out and being on the alert to spot what they're really saying, you make them feel bad and will likely only have addressed one of the several problems they came to you with.
- A useful strategy is for both of you to write down three or four points in advance of the meeting. Once the meeting has started, hand each other your pieces of paper, so you are both responsible for addressing all of the concerns. This ensures nothing is missed, and both parties feel valued.
- Have a notebook with you so you can write down key points. It can help to write a reminder to yourself of how you want to listen. I use 'Listen, reflect, repeat before response.' I also have a stock list of open questions I might insert into the conversation to extract more information.

HOW TO RESOLVE A COMMUNICATION FAILURE

In the times before I started striving to listen to learn, I had many communication failures. Many began with me 'oh-so-cleverly' looking to solve the problems of others. My desire to do so was based on the premise that 'there can be no room for confrontation in the workplace'. These are my words, and I believe in them, but human nature is as it is, and yachts push large personalities into small spaces. There have been many occasions when I have oscillated between mediator, conflict arbitrator and referee. The following story illustrates how even when you think you've got it right, you may actually get it totally wrong, even though the outcome is the resolution you sought to achieve.

Two of the most visible crew members who are key to the yacht's success are the head housekeeper and the lead service. The titles of these roles shift around between yachts, but these two keep the guests smiling, which in turn makes my day that little bit better. On this particular yacht, the crew were in a constant chase to exceed the guests' expectations. Unfortunately, the pace of the owning family's expectations always outstripped the crew's capacity to evolve and meet them. This caused normal and even foreseeable tensions, I have learned with time not to constantly seek to address these tensions; many resolve themselves and just need to be let alone.

As the captain in this environment, the skill comes from learning which issues will and will not settle themselves. I discovered there was a barrier between the head housekeeper and the lead service that was not self-resolving. Both were professional, both were strong, and both would not compromise within their respective departments to ease the burden of the other. With seven days to go before the guests' departure my strategy was to spend time with each crew member daily, acknowledge their concerns (usually about the other), address any simple solutions but not confront the underlying tension until there was time to do so properly.

The seven days fortunately passed without incident and both departments held it together until after the guests had departed. Being an evening departure from the Maldives, I asked to see both crew together the next day during the mid-afternoon. If there was to be any post-meeting negativity, I would prefer it to be late in the day and not infect two teams for a full working day. I chose that we would meet in my office. Normally, I prefer the more relaxed and fresher environment on deck, but this wasn't possible because the deck crew were washing down. My office was well-suited to a three-person meeting. There was a couch, a visitor's chair and my own office chair. In any meeting, I always face away from the desk and turn screens off. I don't like the desk obscuring the person I am speaking with, and I lack self-discipline and know I will continue to monitor messages coming through on the screen if it is in my line of sight.

For this meeting I placed water and two glasses on the desk, as much for the structure of the meeting as for hydration. I wanted there to be a sightline barrier between the two crew – not to obscure, but just enough so they did not have to stare at each other throughout the meeting. I also checked that my office chair and the visitor's chair were at the same height. That provided me with two equal platforms. I would sit on the couch, where I would be lower and, in the middle, the lesser of the three attendees. I was happy with the layout; it was relaxed, but thought had gone into it. I did consider that having one sitting in the office chair and one in the visitor chair might show favour towards one and not the other, but the world isn't perfect, and I was overthinking it.

The two crew arrived within seconds of each other and exactly on time. There was minimal small talk; we knew why we were in the office and like any referee, I laid out the ground rules for our communication. I closed the door and while still standing, delivered my terms: what is said in the office stays in the office; the only way to gain any value is to speak openly – broad, sweeping commentary is less valuable than specific examples; and while I would host and reflect on the conversation, the communication was theirs.

I dropped to the couch and asked the head housekeeper to begin. She opened, predictably, with an indictment of the service department's sense of superiority. As she continued in a monologue, I felt something prodding into my right hip. I was not sitting squarely on the couch and my hip had pushed the cushion aside. I realised that even my disrupting the cushions on the office couch may have upset the head housekeeper's sense of order. She served the same relentless taskmaster I spoke about earlier: her own sense of perfection. Without any interruption to the narrative that was unfolding, I let my hand drop to find what was causing my discomfort. It was a clip-on finger monitor, which must have slipped from the pocket of a technical consultant when he had visited to try and unravel the mystery of a wayward piece of technology. I cupped the little monitor and, maintaining eye contact with both crew, clipped it on to my finger. It is a small device and was easily shielded from view in my lap. I have a low heart rate and when I snuck a glance, I saw my normal 40 beats per minute and 98 per cent oxygen saturation.

With my impromptu health check completed I returned a closer focus to the discussion in front of me. It wasn't going well. Both protagonists were sitting with their backs poker straight and lips pursed. There was a fair amount of 'she did this, she said that' from both sides and I sensed that regardless of my preparations and ground rules, we were just spinning wheels. I asked both to stop and sought to summarise their positions to date. Without being too disparaging, I thought that if they could hear their own grievances reflected to them, they may acknowledge there was little to be really concerned with. I completed my summaries and asked for agreement from both that they were a fair representation. Both gave non-committal agreement, but the body language had not eased, and neither was able to look each other in the eye.

To change how this was going I improvised. I held the finger monitor in my palm, displaying it to the two women, and in a light-hearted tone suggested we use the cheap monitor to see where we were all sitting with regards to receptiveness to discussion. The two crew looked puzzled but due to the authority the position of captain holds, they complied. It was no great surprise that both had heart rates akin to those of an Olympic sprinter crossing the finish line. I drew attention to this and said we were unlikely to make progress while they were in this state.

I asked the lead service to pour them both a glass of water and just take a moment to relax. There was a discernable change after this small interjection. There seemed to be a race to a conclusion, and both gave ground and acknowledged their own role in the tension. The lead service was the more natural communicator and without any prompt, she stood up, thanked me for hosting the meeting, and both women excused themselves and left. She had the finger monitor in her hand throughout her farewell, and as she reached the door, she realised, turned back and placed it on the desk. She then made a comment I thought nothing of at the time: that they would not be needing it again and I should keep it for the next meeting.

In the forthcoming weeks I noticed both women were closer to the point of being friends. When the yacht anchored, they went ashore together for drinks and dinner. I was particularly chuffed with my handling of this; it seemed to have worked out very well and while the little heart rate monitor never made it out of my top drawer again, I thought my use of it had been a clever improvisation.

Some months later the lead service advised she would be moving on. It was a natural time for her to progress in her career, and she had done very well in her role. I requested the purser choose a small restaurant for the farewell dinner – it is better to book an entire restaurant rather than subject other diners to the ruckus caused by 35 crew enjoying free drinks. The two women who had been so at odds with each other months earlier were now arm-in-arm. Emboldened by cocktails, the two were ready to make a point. spokesperson, as always, was the lead service and, true to form, she was blunt in her speech when she told me, 'We hated putting that fucking finger thing on and spent weeks trying to think of ways to punish you for humiliating us.'

I wasn't ready for this and gave the traditional 'deer in headlights response'. The head housekeeper joined in: 'I wanted to do something to your cabin but just couldn't think of what would annoy you most.' I spluttered an apology and said it had been done with the best intentions, albeit I hadn't planned it. Their response was a shared forgiveness; they said they knew

I'd meant no harm and over drinks had agreed that I was just another out-of-touch captain and that at some point in the future they would tell me this. I thanked them for not acting on their instincts and punishing me, and added, 'But look at the two of you now, friends, and now with one departing you will miss each other.' They said that they realised immediately after our meeting that their animosity arose from their similarities and in the weeks following what drew them together most was their animosity towards my flawed leadership.

Key skills

Although in this scenario I made a misstep by using the heart rate monitor, which in review was patronising at best, the rest of the set-up was correct for the situation. Setting an appropriate time, location and seating arrangement is important preparatory work.

Once the parties are in the room, you as the mediator should seek to listen more than you speak, using prompts only where needed to enable both sides to have their say. Take notes. Once you feel the points have been made, it is often worth summarising these and reading them aloud, asking the parties if they agree with what you have noted down. This process of listening back is key to them understanding the other's point of view, and should help you be able to move forwards and find some resolution.

Many times since this encounter I have acknowledged and even used the cohesive force of a 'common enemy' intimacy. I use it to define the landscape when leading a group. The enemy may be the weather, the enormity of our undertaking or an ill-defined third party impacting the team. It has been an important tool to draw upon when motivational cohesion is needed, and to strike down when it manifests against a person or defined outcome. As my mentor and colleague says, 'I don't care if the crew are bitching about me, so long as they are bitching together.'

KEY TAKEAWAYS

- Not all tensions require an intermediary. Some will resolve themselves. The trick is knowing when to step in when to leave it alone.
- Meet in a private place and remove distractions and obstacles that might impede communication. Put some thought into the seating arrangement and try to ensure the two parties don't end up directly opposite each other. Adjust the seating so that they are at the same height, if possible.
- Keep quiet and take notes while grievances are aired, but ensure both parties get an equal say. Read back a summary and check you've got the facts straight.
- Make a note of anything that is agreed to resolve the situation, and again, read back a summary so that everyone agrees with what has been decided.

HOW TO INSPIRE AND MOTIVATE

The last member of the trinity is to motivate. Everyone will have their own way of doing this. You might be the charismatic humourist; if that is who your authentic self is, then I envy you, as you are ahead of most of us already. Maybe leading from the front and just pulling the team along purely by example is your way of motivating. Personally, I ask myself whether what I am doing is likely to motivate others. As I enter the crew mess, am I presenting myself in a way that might lift others? Do I greet others with authentic interest? Do I slow down and give them the time they deserve? This is my unspoken inner dialogue and requires only a moment's unconscious reflection against every action and interaction.

There was a time when I sought to validate my captaincy through the sheer weight of the work I undertook. This statement is made in hindsight: I was not conscious of it in the moment. I would be the first to start, the last to leave and rarely took a break longer than it took to eat a meal. I remember asking the crew when morning and afternoon tea were, since I had never attended. Was this wrong? Was I trying to prove a point and virtue signal my behaviour? No, I don't think so. I had a lot I wanted to achieve and enjoyed what I was doing. Was it motivating for others to work alongside a harried captain who never seemed to stop for a few minutes and enjoy their company? Clearly, no, it was not motivating.

That I fell into this trap is a surprise, since many years earlier it had been observing an overworked naval captain, who refrained from fraternising with crew, that had sown the seed for my resigning from the service. It is therefore a bit ironic that I mirrored his behaviour almost identically a decade later in an environment that was built around having and delivering fun.

Key skills

I will never be the 'happy-go-lucky' captain, but I do now find I have a lot more time to stop and enjoy the company of the crew I work alongside and seek to motivate. There is a place for letting the young deckhand see me having fun and letting my vulnerability show through. I 'say the thing out loud'. I make clear that my going for a swim during lunch break, finishing on time and reading in the sun is my way of showing them that the captain's role is fabulous. This extends to all senior leaders on board: they must sell their roles as aspirational for those they lead. To do anything else is to do a disservice to the future of yachting.

My job is a privilege that I enjoy; that enjoyment does not diminish because I take my responsibility seriously. This is my truth. If it is not your current or future truth, in captaincy or any other leadership role, then you might want to carefully consider if where you are is where you should be. Sorry if this reads as blunt, but if communicating with others all day, listening attentively to their needs and taking their valid concerns as your campaigns do not make you jump to your feet each morning, then leadership is going to be a hard road for you, and it will harder still to lift your teams to their optimum performance.

INSIGHT

Are you a leader? Do you measure your performance from the service you provide to others?

KEY TAKEAWAYS

- There are many ways to motivate, so find a style that is natural to you.
- Pause very briefly before you speak to check that what you are about to say is going to have a positive effect. Try to use body language that matches that positivity.
- Model the benefits of the role you have to others by ring-fencing time for the pleasurable aspects and pointing them out to your crew. If you do not enjoy your job, and it shows, then not only will you put them off progressing, but you'll also find the job of leadership that much harder.

Toolbox 2 – human interactions

The tools in this section are intended to promote thoughts about the practice of leading others, and how to assess and do justice to their value, especially if you are a leader who needs to represent crew in interactions with owners. As ever, these are just my own observations and thoughts on the matter, but my hope is that some concepts and terms presented here can be your guide in a future conversation.

PSYCHOLOGICAL SAFETY

Yachting and the media that surrounds it are rightly smitten with the 'incredibles': the yachts, the toys within and the little technical 'bits' at the cutting edge of design and technology. They are exciting, photograph well and generate significant advertising revenue. I cannot disagree with the attraction of the great yachts. What these great magazine cover photos obscure is the 'human capital' of yachting. I sometimes drop the term into conversations to see what reaction it brings. It goes one of two ways: 1) those in the conversation do not bat an eyelid and keep going as if they know what I am talking about; 2) the opportunists leap on the phrase, as if wondering how they might exploit it. 'What do you mean? Is it an app? Who is leading it?'

I then go on to I talk about a more evolved way to support crew, to improve their standards and lift their lifetime value. About this time, the initial interest often fades, as it becomes clear we are talking about an investment in time and a return that will improve yachting but possibly not the balance sheet. This is understandable: if a business cannot make a profit from an activity, then it is unwise to pursue it. But if there is not a profit motive, what will motivate those holding the purse strings to improve the conditions and performance of crew? Surely, the yacht owners who fund the yachting adventure care about the crew, since they must want their teams to perform at the highest level? After all, their experiences on board and possibly their lives depend upon it.

It would be naive, perhaps embarrassing, to ask an owner, 'Do you want the best levels of service?' The question is rhetorical in nature; we all want the best service. What we all do not agree on is the reality: what are we willing to pay, and how is best service to be measured? The first component is somewhat banal: in our own lives we know we can pay a lot and receive little, and at other times have significant gains from very little investment. Defining what a top level of service feels like, and how much it should cost, is more difficult and there is a temptation to cast it aside by saying, 'It is subjective and variable', but that is to give a pass on doing the work.

When Cornell University collaborated with Google's Project Aristotle to find what was the greatest factor in team performance, the researchers discovered that 'psychological safety' is the most influential, by a long way, when compared with other environmental workplace factors. This was amazing to read, but it was not complete, as the term 'psychological safety' remained amorphous and a bit 'woo woo' for my institutionalised mariner's outlook on life. To break it down to something more tangible, I found a definition within the study that seemed to fit: 'Team psychological safety is defined as a shared belief that the team is safe for interpersonal risk taking. For the most part, this belief tends to be tacitly taken for granted and not given direct attention either by individuals or by the team as a whole.'

'Team psychological safety is defined as a shared belief that the team is safe for interpersonal risk taking. For the most part, this belief tends to be tacitly taken for granted and not given direct attention either by individuals or by the team as a whole.'

This is a great definition, and worthy of slow review, but for what I was seeking to achieve I needed to simplify it and bring it into a yachting context. This is what I arrived at: 'Optimum performance is gained when there is a belief that any crew member will not be punished or humiliated for expressing themselves, speaking up with ideas, questions, concerns or mistakes. They will not be let go from their employment for no given reason and never during their earned leave.'

There was a time when I spoke of mental health, psychological safety and crew development in relation to human capital, all of which sound like costly problems. Now when I speak to a yacht owner or their representative, I instead speak of creating an environment that delivers optimum performance. I lean on the Cornell definition, and my yachting derivation, which lead nicely to me being able to ask that the yacht upholds the obligations in the optimum performance definition.

'Optimum performance is gained when there is a belief that any crew member will not be punished or humiliated for expressing themselves, speaking up with ideas, questions, concerns or mistakes. They will not be let go from their employment for no given reason and never during their earned leave.'

What I realised is that there wasn't a lack of interest in the human capital; the barrier was my lack of ability to frame it in language that made sense to the listener. Likewise, if I couldn't define psychological safety, then I could not implement it. I discovered that providing an environment that cultivates psychological safety was a core value of the highest-performing firms in the world. If I was to lead my small teams to be the best version of themselves, then it was vital I paid heed to this new learning.

Key skills

Some of the core elements around a psychologically safe environment are: trust, respect, understanding of roles, and most importantly, the confidence to be yourself in the workplace. As this fed into my consciousness, it was clear that the teams I had put together in the past, and felt so proud of, were not delivering this at all. We were building teams with cookie-cutter expectations for each role and clipping the edges of creativity. I am not alone in this: there is a tendency in yachting to 'keep your head below the parapet' and 'say nothing lest it placed your job on the line'. This 'just don't rock the boat' mentality leads to decision aversion to the point of stasis. Yachting is an enterprise that by its nature needs to move and interact in a global sphere, so approaching decision-making with fear is many shades of bad – 'bad' being defined as poor decisions across every metric of the yacht's operation. Recruiting, purchasing, maintenance and even the daily operational routines suffer due to this.

I am not overlooking that the captain (leader) is answerable to the yacht owner (the board or ownership structure) and surviving in the workplace is sometimes about just doing what is asked. Stepping outside of the parameters that the employer defines is not always welcome or even possible. This is fine and many a good captain's career is based upon this. However, it comes with a compromise. If any crew members cannot be their authentic selves, they cannot bring their creativity and initiative to the yacht and the yacht will never perform at its optimum.

The word 'yacht' is interchangeable with business, sports team or any group endeavour. That it took a Cornell University study to awaken me to the importance of crew being able to 'be their true selves in the workplace' speaks to how deeply I had been conditioned by my environment(s). Now, I may not always curate the space where all crew can be their truest selves – there is a time and place just to 'do the work' – but with my awareness of the importance of psychological safety for optimum performance, I now embrace and celebrate our differences, and the team performance is richer for it.

KEY TAKEAWAYS

- Consider breaking the cookie-cutter approach to recruiting. Having a more diverse range of people in your crew, who are allowed (within the remit of their role) to be themselves, will improve overall team performance.

- To foster psychological safety, crew need to feel safe to be themselves and to ask questions and raise issues without fear of rebuke.

HOW TO CULTIVATE CREW LIFETIME VALUE

During my quest to learn a better language I encountered the term employee lifetime value (ELV). Substituting 'crew' for 'employee' in conversations, I found once again that I was introducing a term that many colleagues in yachting were unfamiliar with. The holy grail of yacht employment was and remains 'longevity' and I had drunk from the cup myself, when I spoke of captaincies in the past where the crew had remained a long time. This, by unspoken extension, validated my performance as their captain. I saw high turnover as a failing and that my ability to retain crew reflected well on my leadership. This may or may not have been the case, but it suited the narrative I was creating.

The longevity metric is valuable, but not as a stand-alone ideal. Some years before I had encountered the academic concept ELV, I was proposing a new crew's employment terms with an uncomfortably intelligent, insightful and well-educated yacht owner. The terms I was proposing were generous and I did not shy away from this; I knew he demanded high standards. He asked me where these terms sat in the market, and I said in the upper 20 per cent. He absorbed this for a moment, and he responded that he did wish to pay well and was less concerned with the cost than its implications: 'If we pay well then crew will remain, but what if their performance slips or through various reasons they never reach the level we had hoped for at recruitment? They will never leave.'

The yacht owner wanted to know what mechanisms I had considered if crew did not meet the performance expectations for such a high salary. This is where my strong narrative faltered. I think I responded with 'performance review' and 'feedback', but both lacked conviction and, more importantly, missed the detail he was seeking. Through many interests the yacht owner employed many hundreds if not thousands of staff and he knew what he wanted from me. 'Brendan, you need to show me how you will continue to improve the crew's value throughout their employment.' I nodded, but it was only a concept, and he could tell I needed more specifics. 'You need to bring this back to me with a plan for each crew member's development and the performance levels they need to achieve within their roles to both earn the initial salary and to stay competitive for performance-based incentives.' He finished by saying that what he did not want was to have well-paid, underperforming crew who stayed because the terms were too good to leave. If that were the case, he would have to fire the crew to achieve growth, and he did not want to do that.

I left the meeting knowing what was asked but not entirely sure how to deliver it. It took more reading, a few false starts and my seeking counsel from my peers and seniors. I wanted to deliver on what seemed like such a reasonable brief from an informed employer, his time in post-graduate studies at Harvard Business School obviously not wasted.

I succeeded and I failed. The response had a cascading (read complicated) set of benefits depending on position, time on board and ability to influence the guest experience. The pay of the senior team whose performance was best reflected in the performance of the yacht was scaled so that up to 80 per cent of the available bonus was dependent on key performance indicators (KPIs) of the yacht being met. These KPIs included meeting time schedules, budget performance and the most difficult to define, guest satisfaction. The calculation for more junior crew was reversed, with 80 per cent being about their own performance and only 20 per cent being related to the yacht's wider performance. The success was that I met the brief: there was scaling and accountability within the bonus system. The failure was that it remained completely subjective in the end, as there are no metrics such as sales or production targets on a yacht to measure against.

Key skills

If I was to tackle ELV again and deliver on a client request I would reverse the flow. My approach is now to ask every crew member during the interview process three questions that I will repeat throughout their employment tenure:

1. What is your potential contribution to the yacht across three areas:
   I.   core professional skills application
   II. crew culture
   III. the guests' experience?
2. In your past experiences, what barriers have there been that stopped you contributing fully?
3. What do you need from me, from the yacht's leadership team, to remove these barriers?

This is an accountability pact between me as the captain and the new crew member. Quarterly or as needed from their commencement these responses will be reviewed. It is a 'feedback, feed-forward' agreement, unlike the traditional binary feedback conversation. The crew member makes a statement about their potential and backs this up with examples of where they have been held back in the past, and the captain agrees to (where possible) remove barriers and facilitate their growth so they can reach their fullest potential.

When this works, the crew member is unlocked and far more capable of delivering their long-term value. If it was a lack of formal training certification or experience in a certain area that was limiting them in their past, then once this has been redressed, they restate their potential contribution and a new performance contract is entered into. I use the training example because it is the easiest, but what I commonly hear is, 'I did not reach my potential because the captain shut down my ideas, my voice, my contribution.' I nod with my commitment not to make the same errors as their past captain (my cohort), but at times I know I will, and working on this is part of my own lifetime value plan. I may not need to complete the next level of training, but I do need to allow my team to feel safe and contribute as their true selves, so they can achieve psychological safety.

KEY TAKEAWAYS

- Crew lifetime value is a continuum, not a percentage annual salary increase. It is a shared contract to perform between the crew member and the yacht's leadership and management team. For this to work, all parties must be held accountable.
- Employee lifetime values (ELVs) need to be reviewed regularly, and barriers to higher growth need to be identified and ideally removed.

HOW TO GIVE AND RECEIVE FEEDBACK

When the founder of Netflix writes a book, read it. When the CEO of Google hands you a book (that he wrote) about the person he respects most, read it. I dutifully read *No Rules Rules: Netflix and the Culture of Reinvention* and *Trillion Dollar Coach: The Leadership Playbook of Silicon Valley's Bill Campbell*. The thread that ran through them was that open and honest feedback and candour is a cornerstone of their shared success. Both books provide frameworks for this and caution against the temptation to be the 'blind cheerleader'.

This struck a chord: it was something I fell into the trap of doing. I enjoyed telling crew when they had done a good job but struggled to give constructive and real guidance when there were observed shortcomings. The praise was received well, but crew were seeking insight into how they could grow. To give only loose praise – the metaphorical high five – was weak and played on my wish to always be the pleaser.

For candid feedback to work it must be given in all directions. There must be a culture where it is clearly stated that crew can give feedback upwards, peer-to-peer and to those they lead. The value of the communication does not change with the direction of the relationship. Some of my greatest personal growth, such as the time when the lead service and lead housekeeper told me about their joint dislike of my use of the heart rate monitor, came from 'truth to power' feedback from the crew I sought to lead.

Some years ago, a bosun I was working alongside who has now continued to a great captain's career of his own, said to me with no malice, 'Brendan, you talk at me, not with me. It is not enjoyable.' Never before had I received a more succinct wake-up call, and it has shaped my communications since. I was lucky to have had a truth-teller, confident in their own position as a bosun. Without his intervention I may have gone through my entire career without learning of my flaw. To remove the element of luck, though, there is structure that will assist the less confident, the culturally deferential, in speaking out.

Key skills

The excerpts below are taken from Netflix's playbook, as presented in the book, *No Rules Rules*. Within Netflix the mnemonic '4As' is used to quickly recall the words: assist, action, appreciate, accept or discard.

Giving feedback
Aim to assist:
Feedback must be given with positive intent. Giving feedback to get frustration off your chest, intentionally hurting the other person, or furthering your political agenda is not tolerated. Clearly explain how a specific behaviour change will help the individual or the company, not how it will help you.

Actionable:
Your feedback must focus on what the recipient can do differently.

Receiving feedback

Appreciate:

Natural human inclination is to provide a defence or excuse when receiving criticism; we all reflexively seek to protect our egos and reputation. When you receive feedback, you need to fight this natural reaction and instead ask yourself, 'How can I show appreciation for this feedback by listening carefully, considering the message with an open mind, and becoming neither defensive nor angry?'

Accept or discard:

You are required to listen and consider all feedback provided. You are not required to follow it. Say 'thank you' with sincerity. But both you and the provider must understand that the decision to react to the feedback is entirely up to the recipient.

Adapting a global tech company's guideline into a yachting culture needs some finesse. I expect that if you are a leader then you will have enough cultural awareness of your workplace to make this adaptation. One additional tool is to build a code, a shared language that can be used to pass an important, yet uncomfortable message. 'Mildred' is an example of how this was done on one yacht I worked within and demonstrates how powerful a tool shared language can be.

'Mildred'

I had a close professional relationship with the chief officer; we shared values and professionally enjoyed pushing each other with the diversity in our knowledge. In hindsight, while the friendship was true, I did utilise my position to (at times) talk down and mock my friend and colleague. As professionally and socially competent as the chief officer was, when he was fatigued, he moped and saw the negative in everything. This is something we are all guilty of; I just happened to observe it in his actions and since it is one of my own flaws, it triggered me.

Ironically, his fatigue was most often induced by me pushing him and the crew beyond what was reasonable. Nevertheless, in one of these moods I said to him, 'You're like Mildred.' He asked what I meant, and I went on: 'You know, the English middle-aged wife character from the '80s television show *George and Mildred,* nagging all day and seeing the worst of everything.' He gave a 'laugh' – an 'I wish Brendan would just leave me alone' laugh.

I revelled in my humour, albeit laughing alone, but nevertheless 'Mildred' became a reference for anyone whose bright light dimmed through fatigue or lack of perspective. It was a top-down comment, or at least I thought it was until maybe a week later when I entered the crew mess and saw on the lunch-menu whiteboard a well-drawn captain's peaked cap with the words 'Mildred Alert' placed above it.

My words had been used against me. Well, not really against me, maybe even for me. One of the junior crew had heard through the yacht that I was tired, behaving poorly and generally just needed to be called out for it. The code word 'Mildred' was the perfect way to express this observation and as I read it, I looked around and realised everyone was watching for my reaction. I put my hands in the air, faced the room, smiled and said, 'I get it: I am being a dick. Thank you to whoever wrote on the board.'

From then on, 'Mildred' became our code for giving feedback that behaviour had slipped. It was so much easier to say, write and share than to have to go through the tough conversation that was needed to call each other out for poor behaviour. Giving crew the chance to draw attention to otherwise awkward topics using mutually agreed 'codes' is powerful. It is empowering to not have to use confronting terms and so long as the leadership allows themselves to be accountable to their use, then it is a tool I highly recommend.

KEY TAKEAWAYS

- For candid feedback to work it must be given in all directions. Many of the greatest insights come from people feeling able to speak truth to power to their leaders.
- How you give and receive that feedback is important:

  1. It should be given out of positive intent: to benefit the individual or the company, not you.
  2. It should be clear and actionable, so the recipient goes away with a road map of what is required.
  3. When receiving feedback, try to be appreciative. Listen with an open mind and try to avoid becoming defensive.
  4. You do need to listen to feedback, but you do not have to follow it. How you react is up to you.

Having a code word that can draw attention to poor behaviour without being too confronting can be a useful tool, and can keep an important topic light.

HOW TO IDENTIFY AND USE VALUES AND BELIEFS

To communicate effectively, you need to know what you believe in and why. By examining the filters that you pass all information through, especially when making a judgement call, you can better understand why you respond in the way you do, and potentially change that if required. To do this, we need to examine two terms that are ubiquitous both in the business world and in media in general: values and beliefs.

Values

Values, values, values. Everyone was speaking about their values, and I didn't like it. I didn't like it because I was conscious that I had not and could not clearly define my values, either personally or professionally. I didn't even know if they should be the same or different. I felt I was living a good life and there was a guiding thread, but words matter, and if I couldn't communicate to others what my values were, then I would be limited in how I could bring them into my life.

Mixed in for added confusion were beliefs. These seemed to overlap in this consciousness soup, and I didn't stand a much better chance at defining them either. Was there even a difference between values and beliefs?

My meandering thoughts came into sharp focus when I was listening to Jordan Wylie during one of the many Zoom 'get-togethers' during Covid-19 times in mid-2020. Jordan was a former marine, published author, supporter of multiple charities and all-round good person to listen to. His very uplifting talk had finished and there was a question from the online audience: 'Do you have any structure to help in making critical decisions?' Jordan did not skip a beat and replied, 'I know my values and I pass important decisions through my values as a filter. I know by doing this that the output will always be the right decision for me.'

Thanks a lot, Jordan, I thought. Just as I am struggling to communicate my own values you provide such a perfect example of their importance to life and to leadership. My timeline to work out what values mean sped up at that moment. Soon after the call with Jordan I accepted a podcast interview request from the wonderful founders of the diversity in yachting organisation SheoftheSea. Their message is important and as I said in Chapter 5, the definition of diversity needs to be widened to bring yachting in line with modern expectations and give it a chance to perform at its optimum. The interview should appear fresh and live, but for the sake of quality I wanted some chance to prepare, so I asked for the organisation's questions in advance. I could have, should have, foreseen that Jen and Natasha of SheoftheSea would have in the question list: 'What are your values and how do you apply them in your workplace?'

There seemed to be no escape, so I started to make my list. I placed a caveat against them, stating that these are not what I achieve each day; they are the values I aspire to. Here is that list:

- Grit/perseverance – I get up and go again in the face of adversity.
- Trustworthiness – financially, personal behaviours, safety.
- Integrity – I stand by what I do, even if it is at my expense.
- Energy – I live a healthy life and my return on investment (ROI) for this is having energy each day.
- Efficiency – I look to do everything as efficiently as possible. This can be a curse.
- Adaptability – I work well in disruption.
- Self-reliance – I don't expect others to solve the challenges I am confronted by and I take ownership for my situation. I do seek help but not to carry my water.
- Curiosity – learn … repeat.
- Positivity – I love yachting with all its wonder and its problems.
- Humility – Life is not about you. It's about what you do for others.

It sounded good, but it is a long list and if I used it in the manner Jordan had described, then I had put too many filters in line and just like passing anything through a filter, if the filter is too complex, nothing will come out. Time passed and I cannot even recall the trigger for the rethink, but I eventually identified a shorter set of values, which without reducing the importance of my first attempt, were less about virtue signalling and more about my own truth. However, there was one that I had missed the first time that was truly fundamental to the point that when I observed its absence in the list above, it upset me. It is an unconscious value that was going to be the first filter for my future decision-making: kindness.

Kindness
Here, I define kindness as: kindness in my interactions with the world and observed kindness from others interacting with their worlds. Their kindness may be towards me, towards a colleague or towards the person serving at a restaurant – all are of equal merit. I realised kindness was a value I held so highly that observing its absence, even when I was not involved, made an impact on me.

I cannot control the interactions of others, and do not seek to, but the realisation of its impact meant I had finally identified the importance of values in decision-making.

My goal is to use kindness in a way that contributes, to be a giver of energy and not a taker. I live in a rural area and communicate in my second language. I therefore rely on the patience and understanding of others to have efficient communications. An act of kindness is as simple as waving to someone with a smile when you know your communication may be limited by language. When paying for groceries or ringing a call centre, I want to take that moment to care for the other person. Those who know me may smirk and point to my failure to consistently achieve what I have just written, but to them I say that the opening words are the most important: this is my goal, and as easy as it is to write it, I do not always live it. When things are (by my own definition) not playing in my favour, my tone can become curt and I hold the person accountable for the emotion I am feeling at the time, even though they have no chance of even knowing my feelings. I acknowledge this flaw.

By saying out loud that kindness is my highest value I am making myself wholly accountable. I am ready to be called out when I fail. It may seem a tenuous link, but I think my intention exactly mirrors a navigation passage plan: by stating what the goal is to others, I am automatically making myself accountable if I deviate from it. One of my attractions to endurance events is that I do not begin knowing if I will reach the finish line. The same is true of my goals for the values I aspire to. If they were easy to achieve all the time, what would be the challenge or the benefit?

Humour

Trailing closely behind and directly linked to kindness is humour. Life's just too short not to make a joke when it can be made. The inner child sometimes gets hidden or forgotten in the workplace. I look to find and release this side of me. There were years when I did not pay attention to this and I lost contact with my playful self, thinking seriousness was the key to success. Fortunately, I crossed paths with some very successful 'children wearing an adult's shell' and realised my seriousness was not true to myself and further, not needed. This is not in tension with taking my work seriously. I do not foster an environment where flippancy and poor preparation may cost lives. Appropriate humour layered over professional preparation and execution enhances, not detracts from, delivering a professional and safe operation.

Grit

My nephew stayed with me and my family for a while some years back. We came to an arrangement where he would come to Austria, live with us, work around our house and I would pay him in Australian dollars to give him a head start for his life at home. He is a nice young man but lacked a bit of … well, let's just say application to a task. Each day my wife and I would think of what we should say to motivate him and finally I whittled it down to 'Don't give up.' Whether you're painting a fence, walking up a mountain or learning to ski – just stay with it. It does not need to be some incredibly audacious goal, either; perseverance at the granular level is equally if not more valuable.

One particular day, I had asked my nephew to move soil into a raised garden bed with a shovel. He seemed to pause after several cycles and reflect on what he had done. He was not fatigued or in pain; he was just looking for some way to cease the work. We all do this in our days and sometimes the refrain of 'just keep shovelling soil' can be a metaphor for everything we place as barriers. In Steven Pressfield's classic *The War of Art*, among many truly life-changing insights is the notion of 'resistance' as a force against creative achievement. 'Resistance' is a character in the book and has a life-force that pushes back against our chance to achieve all that we can. It is a shape-shifter that will impede given any opportunity.

The counterbalance to resistance is grit. By publicly stating grit as a value, you are placing an obligation on yourself to not give over to resistance in its many forms. I stress here that grit does not relate to working through pain and physical injury. Grit as I like to use it in my leadership and personal values is to assign focus to the task and to acknowledge that Pressfield's 'resistance' is ever-present and that I will demand of myself the strength to stay true to a core focus.

Respect

I respect myself through how I live (broken down to diet, health, behaviours) and from this base I feel I am in a good place to respect others for their similarities and differences. I long ago had my 'Copernicus awakening' – that is, I learned that the world is not revolving around me and events occur not *to* me, but due to their own nature and merely catch me in their path. I respect the fact that we all make mistakes – individuals, companies, town councils and nations. I think more often this is just due to human incompetence than any conspiracy and certainly not one directed to me. This releases me to feel full agency for everything that happens to me – all the good, all the bad and all the ugly that occurs in my life is by my own actions.

When I talk to teams I lead I speak of the cascade of respect. On the inputs side it begins with the base layer (self). The next tier is others in your orbit (colleagues), then comes the environment you are working within. On the output side there is respect for the client/customer. In many business practices it is tempting to put the output side as the priority, but this does not work and is a cause of so many organisations not achieving their goals in delivering their product or service.

Efficiency and energy

Efficiency of thought, efficiency of action, efficiency of emotion. Take Occam's razor to life and don't make anything more complicated that it needs to be. In a recent meeting with a semi-retired, self-made billionaire he repeated the phrase (to the point of awkwardness): 'Why can't your team understand, I don't want things complicated? I just want the simple things done well.' This conversation was specifically about the yacht crew's support of his requests and is a problem many will never encounter, but the value itself is worth pursuing: approach tasks with an elegance of simplicity.

This is a theme explored by John Brockman when he engaged some of the world's greatest thinkers to create the fabulous book *This Explains Everything: Deep, Beautiful, and Elegant Theories of How the World Works*. Not surprisingly, the idea that was most recognised for its beauty by these stellar polymaths was Charles Darwin's theory of natural selection. Darwin's world was full of complexity, noisily amplified by religious zeal and mercantile prejudice, yet he was able to cut through this and propose the simplest and most efficient theory that addressed biological history's greatest questions. Few of us are as brilliant as Darwin, of course, but we can still strive to apply this principle to our work.

Beliefs

Beliefs and values are the same thing, right? Not at all. I believe the sun will rise in the east and set in the west and I believe this intrinsically. But, if the sun reversed its relative path, I would change my belief accordingly. My beliefs are based upon the verifiable knowledge I have today, and I am open to changing them when presented with a stronger set of facts. I grew up with Pluto as a planet, and while I mourn the loss, I do now believe there are eight planets in our solar system.

There was a time when I applauded someone who presented with a strong set of beliefs; I thought it showed maturity of mind. I no longer view it with such clarity. I think all beliefs should be tempered by an awareness that they are subject to review. This is the difference between beliefs and values, and I think it is important in my captaincy to define and acknowledge that difference. My values may shift in weight due to circumstance, but having taken the effort to interrogate what really matters to me, they are now fixed into the 'who I am category'. My beliefs, on the other hand, are wide open to being overturned when better information presents.

KEY TAKEAWAYS

- Beliefs and values are different things. Beliefs are ideas and opinions based on verifiable facts and can (and should) change according to the latest information. Values are based on what is important to us – how we attribute worth to objects and behaviours.
- It can be hard to pinpoint your values, but it is an exercise that is worth doing, especially if you are in a position of leadership. Start by writing down everything you can think of and then aim to distil this to a shorter list of essential filters that you can pass your judgements through.
- Values such as kindness, humour, grit, respect, efficiency and energy are both general and specific, and can be brought to bear in any situation. Those are my values, though, so take the time to work out what yours are.
- Respect can be broken down to inputs (respect for yourself, those in your orbit/colleagues and the environment) and outputs (respect for clients). Although outputs are often prioritised in businesses, it is in fact the inputs that should be the priority.

THINKING BETTER

Learning my values and then being able to use them as a filter while I was communicating gave me a start, but I knew that captaincy would demand that I made many complex decisions and that values alone might not be enough to improve my communication skills. It intrigued and then annoyed me that I would be required to make decisions all day, every day, yet I had not received any education on how to do this better. I began to try and bridge this knowledge gap for myself by turning to research in the hope that I would then understand how and why I made decisions. Understanding this would improve my work performance in general, and in extreme situations my ability to make better decisions might save or cost lives at sea.

My first step was to examine the following factors:

1. How many people does my decision affect?
2. Is it time critical?
3. Do I have access to counsel on this or must I face it alone?
4. Is it reversable?

From this I began to look to how I could group decisions into categories to help me. I looked around for someone else's solution; there were many to choose from, but none really spoke my language, so I patchworked together my own solution by drawing on ideas from a few different places.

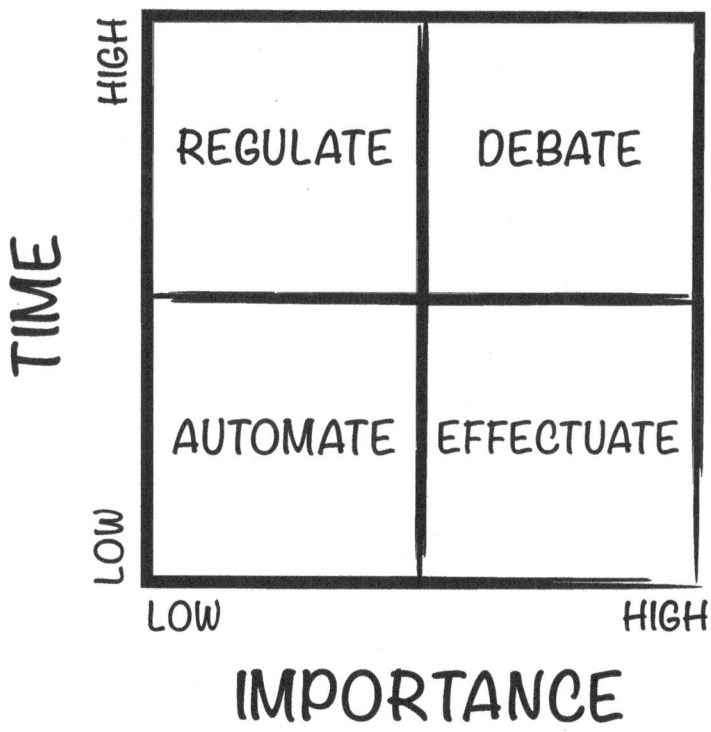

NEIL PASRICHA'S DECISION MATRIX

Among those influences is Neil Pasricha and his decision matrix. This is a map to place decisions into, a way of parsing information into manageable, predetermined approaches. After hearing him explain his humble decision matrix, I discovered it works for me and I try and factor it into my days. I wouldn't say I am driven by it, but it does help. The awareness that not all decisions are critical stops me from staring in paralysis at the lunch buffet and frees the time to focus on those decisions that are significant. It also encourages me to ask a few vital questions:

- Can I automate the low-importance, low-time decisions, to free up time for making more complex decisions?
- Can I regulate the low-importance, high-time decisions (the 'time vampires') to allow more time to dive deep and debate high-importance, high-time decisions?

A doctor of medicine is required to make decisions and execute a strategy when a patient presents with a challenging case. They will look to reference sources where similar cases are recorded and review successful strategies to address the situation. They might not stop there: they may use peer-reviewed papers to discover even more ideas and then refer the patient to a specialist in exactly that area to round out their strategy for the patient's treatment.

Sadly, there is no reference book published to guide the decisions I have to make, but I did need one. So I wrote the guide myself. It reflected by omission the knowledge I already held and, more importantly, covered what I was lacking. These two categories shift backwards and forwards sometimes and the toolbox of strategies to make decisions shifts with them. I had been doing this for some time and felt validated when I saw a quote from the great businessman and philanthropist Charlie Munger:

> Well, the first rule is that you can't really know anything if you just remember isolated facts and try and bang 'em back. If the facts don't hang together on a latticework of theory, you don't

have them in a usable form. You've got to have models in your head. And you've got to array your experience both vicarious and direct on this latticework of models. You may have noticed students who just try to remember and pound back what is remembered. Well, they fail in school and in life. You've got to hang experience on a latticework of models in your head.

I have heard myself and my peers say when pressed, 'Look, I have been doing this for xx years and I know what I am doing.' Time on task is being used here to establish a defensive position, digging in a competence trench to repel challenges to accepted views. Over time, I have worked to avoid this in my own narrative and my ear is more tuned in when I hear others leaning on experience as their crutch. Yes, in my core competence area, time on the job does assist – my thousands of days at sea give me an arsenal of solutions based on past events – but it is not my leading skill and attribute.

Time is not enough. Yes, many events do have the same beat and experience helps, but what if my past actions were flawed and any successes in that time were not correlated directly to my behaviours? If I believe my actions alone were responsible for an outcome, I am denying that there are many other influences beyond my control. Even stronger is that the favourable outcome, while achieved, may even have occurred *despite* my actions. The forces beyond my control may have been so strong that the outcome was predestined irrespective of my input.

Regardless, leaning on time on the job without updating my knowledge and approach to challenges will result in me repeating flawed behaviours. It will lead me to reinforce poor decisions until I am so enamoured with my choices that I am closed off to alternate solutions. As Munger so simply points out, I need models to work with, to hang my problems on so I can see them flow through to resolution.

Having recognised that I was not updating my knowledge and that I was wandering aimlessly through the defensive trenches of experience alone, I sought to overcome this. I had moved comically along a well-trodden path where I grew through my imposter complex and enjoyed a moment of calm before realising I was now exposed to the 'competency trap'. This is a trap that held firm the belief that my experience and past success remained valid and would ensure success into the future – just play the same tune and the crowd will keep dancing. This doesn't work and I was lucky to see this in others, as I may not otherwise have recognised it in myself. Just like the tools involved in my trade have evolved from sextants to satellites to Starlink, I too have had to make sure my decision-making is not static. I have therefore undertaken a quest to learn appropriate strategies to help me make decisions. The strategy must be appropriate to the situation and, importantly, I have to understand and believe in its application.

Key skills

I know I cannot recall all that I read or listen to. I have already introduced my process to continue my learning and I will quickly repeat its simple and humble pattern: I highlight passages that I wish to retain, I rewrite these in a notebook I always have with me, and then I thumb this notebook and scribble again in columns next to the good bits.

Curating what to read is more challenging. The only guide I can give is that you make the act of reading a daily habit. Not everything you read is going to be gold, but increasing the volume of what you read will increase the chance of you sifting through the noise and learning in pace with experience. There is an accepted idiom that says you become the product of the five people you spend most time with. I extrapolate this to say your views are a product of the most recent 50 books you have read.

INSIGHT

I know that all I know today will be invalid soon and my learning wheel must keep rolling.

KEY TAKEAWAYS

- Making better decisions requires granular and first principles thinking. You need to ask a cascade of questions, including determining how critical in terms of both time and importance a task is. Some tasks can be automated, which frees up time to focus on the more complex decisions.
- Good decision-making requires a strong framework, not just the recall of past actions or facts. You cannot rely on past decisions alone to inform those you make in the present. Your knowledge and skills need to advance through a process of constant growth and learning.
- Highlighting passages in books and writing them out in a notebook, serves both as a quick reference, and will embed them better in your memory.
- My method for continual progress has become read, learn, practise, grow, repeat.

MAINTAINING AND IMPROVING STANDARDS

Reporting directly to yacht owners has taught me that offering an excuse, regardless of its validity, is not appreciated and it is easier just to accept the chastisement and move forwards. On the rarest occasions when expectations are met, the yacht owner will automatically and unconsciously shift the target higher again. My life as a yacht captain is therefore never about meeting static expectations and resting back in satisfaction. In fact, on the rare occasions when I have met all expectations, I have become even more alert. Like being out in front in a race, I didn't know what was coming behind me; at least when I was defending performance that was deemed subpar I knew where I stood.

The only way I have ever had a chance of meeting such shifting and ever-ascending goals is to work under the guidance of a vision. I didn't quite know the importance of this until I saw a vision statement at my friend's business in Australia. I read it and saw that it defined the entire 'why' of the business, for the staff and for the customers.

All the yachts I had worked within were striving for their form of excellence without ever determining a vision. It is the nautical equivalent of putting the engines to full ahead without a navigation plan, and then going to lunch. I was among those who did this: I approached captaincy with great horsepower but with not the greatest control. It was not until I was a long way into my career and after an epiphany in an unrelated business that I understood the power of a statement of purpose, of the intention to lead a team and manage the process.

I did not need to be the sole creator of the vision – quite the opposite: the more stakeholders I could involve, the more chance there was the vision would be followed. I knew as captain that I should be the shepherd to guide the vision through to its completion, but I would need to seek and then embrace direction from the yacht's owners, their representatives and their appointed managers. The needs and wants of the crew would also need to be married to the vision without there being too much compromise. Yes, this is a needle to be threaded, but I thought the investment was worth it.

Timing is everything, so one day, during the construction of a new yacht when there was a great feeling among the crew and full engagement from the owner I had an 'If not now, then when?' moment. Many of us had worked together in the past and we were supported by an incredible owner who had a clear sense of style and purpose for his new yacht. His personal why and intentions were very clear. We were all seated for morning tea in the converted shipping container we called our office in the ramshackle shipyard where the yacht was being built. When I rhetorically asked the team if there was something more we could do to glue our performance together – something we would refer to in times of need and reward ourselves with in times of success. What did we want to be a part of? How did we want to be seen by others? How did we want to see ourselves and how could we grow the team, yet retain the consistency of what we felt we had brought with us?

In hindsight, it was a bit much to launch unannounced in a morning tea session and I saw more blank faces than engagement, but we moved forwards and what started as a loose collection of ideas started to form and gain momentum. It seemed we all wanted to be a part of something; we did not want to put the engines to full and sail away with no shared plan. The list grew, shrank and grew again as we all contributed and self-edited.

Around the same time, a leading industry magazine launched a new publication, *The Superyacht Owner*. The editor contacted me and asked if there was anything we wanted to contribute. Our project to curate a vision, it seemed, was relevant to their audience. It was flattering to receive their request, but the crew were hesitant. The vision we had come up with was seen as something quite personal and not to be broadcast. I remember campaigning that since we were proud of what we had achieved we had a responsibility to share the message. In the end, we found a middle ground: we did publish, but we shared anonymously.

Key skills

We had built the vision by identifying and breaking out the key elements. It wasn't some dissertation on management, but the real words we used each day. We asked some probing questions along the way:

- The big picture: Why were we going to work each day? What were the core metrics that underwrote our performance?
- Brand: Our brand was not a logo. The brand needed to embody the values of the yacht and go further to guide behaviours and interactions. We needed to support the brand, and the brand had to represent us.

- People and personal values: This was big – how would we interact with each other and with the wider community our yacht moved within? What were the core values that we needed to have as a team?
- Leadership: We talked about what style of leadership was going to be in place. How were the seniors going to lead and how were the juniors going to interact and challenge healthily? This drew the most interesting response: there was a fair consensus of what style we did *not* want but it took longer to identify what style we could uphold.
- Communication: Within the vision it is hard to choose a 'this is most important', but agreeing on how information that underpins the operation is conveyed was essential. Only with agreed pathways could our information and its content be legitimised.
- Measuring and monitoring: Everything we hoped would be achieved would only stay valid if it was reviewed for its relevance. It would only be embraced when the team saw that the vision improved their performance, their engagement and their enjoyment, and this could only occur through pre-agreed measurable outcomes.
- Success: As with measuring and monitoring, defining what success looks like was just as important. If I or any of the team couldn't see what success looks like, there was no reason to put the effort in. An Olympic athlete training through heat and cold for four years remains driven by the chance to cross the line first. This is easily visualised and sustains their discipline through the hard hours of training. What was our visualisation of success? Did it vary between crew members?

After many drafts we shared the vision by framing it and hanging it in the crew canteen. This then became the crew mess once the crew were able to move on board the newly finished yacht. It was a visible representation of our commitment. To take it further and to embed the vision as our intention, it became part of recruitment and induction. We shared it with new candidates and asked them what they understood from it. There was no wrong answer, though it was important they could identify with the reason for the vision and ultimately, they could describe what their success in the workplace looked like. If they could not see their own success, like the Olympian crossing the line, they would not commit the effort to achieving it and therefore were likely not the right fit.

The yacht owner, their representative and their appointed managers were all stakeholders and held copies of the vision. Indeed, the yacht-owning family's intention on starting the yacht construction was the inception, so they needed to be shown the evolution and the crew's embrace of their original ideas. It was not lost on us that it put us on notice: the yacht owner and their external stakeholders held us accountable to deliver on the goals of the vision. I had many darker days when my or the team's performance did not meet the standards of the vision. The only thing that helped was knowing that had there been no vision, I would not even have known that I was off-track. Just like my learning in pilotage, without the plan, the yacht does not know if it is standing into danger and the faster it goes, the bigger the crash at the end. I think without our vision we would have had more and heavier crashes.

KEY TAKEAWAYS

- Yacht owners' expectations are constantly shifting and increasing, meaning it is very hard to meet them. One way of handling this is to have a vision: a mission statement that defines the why of the work, for the crew, for the owner and for you.
- The vision needs to be defined by many people in order for it to work, including the owner, the owner's representatives

and their appointed managers, with some crew input. If you are a captain, it is your job to guide this process.

- To create the vision, you need to break out the key elements, then define their meaning and how they are applicable and can be measured. Use everyday language that is easily understood and is relevant.
- To embed the vision it needs to form part of recruitment and induction processes, so that new crew are aware of the vision from the very first moment.

Loose guidelines, tight objectives

Having worked hard to define, implement and share the vision I needed the day-to-day strategies to achieve the mission it created. If the vision was where we were going, then the mission was why we were going there. The mission needed to be communicated with clarity and simplicity, and I struggled to work out how to do this. Once again, I was assisted by a close friend with no attachment to the yachting industry.

I first heard the term 'loose guidelines, tight objectives' during an afternoon walk with a friend. Doug is a natural communicator and a fun guy to be around; he was correctly placed in sales. He had just returned from a fundraising lunch, the speaker of which was an eccentric character I had heard of but never quite looked into in detail.

Colonel David Hackworth was a controversial former American serviceman who had retired to Australia and now spent his time writing and speaking of his military experiences. During the lunch, Hackworth had described a situation during the Vietnam War where, as commanding officer of the Ninth Infantry Division, he and his men were being pinned down. He had drawn his company commanders together and delivered his mission brief while pointing at the field map. According to Doug, Hackworth's speech went along the lines of:

We are pinned: the enemy are surrounding us here and here. We must make it to this feature in two days to support a planned major offensive that when successful should shorten this god-forsaken war. You are all great leaders, and your men are superior to the enemy, you will make it. Check your watches. We meet again in 48 hours.

The mission objective is clear, as is the time frame. Yes, the field of battle may present a clear objective, but I was already approaching every day on board as a battle to improve my own leadership and often think that the 'fog of war' demands and delivers the finest leadership examples, even when in support of not-so-fine pursuits. In this case, Colonel Hackworth had clearly stated the objective and, more importantly, made it clear he believed his teams could achieve it.

As I listened to Doug, I thought how it must have felt to be scared for yourself and for the men in your command in that Vietnam jungle. I would have needed to know why I was putting lives at risk and that the men I was leading believed in me and my team's ability. Colonel Hackworth also provided guidelines; they were brief. He then simply told his company commanders to use the training he had been brutal in delivering for the past six months and to not die themselves or kill their men by taking shortcuts.

It could have been very easy for Hackworth to have laid out to each company commander how they should move their men and how they should engage the enemy. This is the sort of top-down instruction I had received in so many of my workplaces and had been guilty of doing myself, telling my teams exactly what they should do each day, checking on their progress and then praising or chastising their ability to complete the assigned tasks. I never questioned that through this process I had removed my crew's ability to use their skills and abilities with freedom and, more importantly, to know *why* they should labour hard to complete their work. In other words, there had been no objective.

Colonel David Hackworth and his team did indeed achieve their objective and their guidelines were maintained: there was no loss of life. History cannot be rewritten, but when Doug had finished retelling the story I wondered what may have happened if Hackworth had approached his briefing as a manager would, giving tight guidelines and skipping the objective, rather than as the true leader he was.

This story stuck with me and I turned it over in my mind many times to gain its benefit. I integrated the concept of sharing and tightly defining the objective, yet allowing those I led the freedom of not overly defining the guidance. It was not a natural shift and went against some learned behaviours. Yachting was all about process and tight guidance, so how could I let this go? Would my crew know how to do what I asked if I did not tell them?

It seems that the crew knew more than I ever gave them credit for and, often without my overbearance, came up with better strategies than I would have done to achieve their mission objective. It was an awakening. I was one step closer to becoming the captain I hoped to be.

As a footnote: Hackworth's ire was awakened when it became clear the mission objective, where he risked the lives of his men, was indeed flawed and there was no early end to the war in sight. In a time of tight conformity, it was unheard of for a uniformed colonel to speak out on national television, yet that is what he did. Taking his eight purple hearts and ten silver stars for gallantry and bravery with him, Hackworth retired early after a Pentagon investigation into every aspect of his life found no wrongdoing.

COLONEL DAVID HÁCKWORTH

KEY TAKEAWAYS

- The vision is where you are going. The mission is why you are going there. Both need to be communicated with clarity and simplicity.
- The concept of 'loose guidelines, tight objectives' involves providing your teams with a clear objective, within defined parameters, and an explanation of why it needs to be done, but then trusting them to use their training to achieve it. Rather than micromanaging every decision and action they take, you give them space to find their way to the objective themselves.

183

HOW TO COMMUNICATE AS A LEADER

The unorthodox, yet incredible, military leader David Hackworth's leadership mantra was a great fit for captaincy and it is not the only military lesson I draw upon. The military may mean different things to different people, but they have always known the need for leadership, and even sometimes without knowing, have led the genre.

Yes, OK. The military may have had a rocky start with indentured service, lashings, beatings and much, much more, but they have come a long way and in my short exposure I was able to learn some very basic, yet completely transferrable leadership lessons. The points I can offer from these years in military service are:

1. Learn your people.
2. Be friendly, but you are not their friend.
3. Sir always eats last.
4. Don't run: it will scare the troops.

This is all well and good, but what do these mean in practice?

'Learn your people'

It may seem obvious, but remarkably few leaders invest in really knowing the people they lead. This is likely because it takes time, does not give immediate rewards and there isn't one single way to do it.

I maintain a notepad and jot notes from crew conversation. How old are their children? What was the last big event they wanted to get home for? Do they have pets? Are they musical? What are their interests beyond the yacht that are not normally visible to others? I find by writing these notes by hand, I retain the information and can recall it easily at a later date.

'Be friendly, but you are not their friend'

As a junior officer especially if you are young, it is very tempting to seek the company of junior sailors who are more similar in age to you and more welcoming as a social group. Of course, there is nothing wrong with this and rank should not be a barrier to friendships, however, leadership only works if there is respect, and nothing erodes this faster than a lapse in social behaviour.

I know I wrestled for the longest time to find my social place and strike the right balance in trying to lead while still being approachable. Too many of my colleagues showed me the example of what not to do and damaged their leadership through inappropriate behaviour, and I am not without guilt by this measure. I knew that as a superyacht captain my actions were under greater scrutiny than those of everyone I led. This is not unfair and reflects community expectations for leaders.

In practice, this means that if there is an alcohol-related event you are expected to attend, you should mingle, contribute and depart in the same state you started. You are not the friend. You can be cheeky and join the banter with humour (see p. 166), but never forget that your jokes or innuendo-based comments will cross a line from being funny to inappropriate in an instant – the instant being defined by the crew in a flash before it is visible to you as the leader.

This is easy to write and hard to live on a yacht. One reason yachting is so appealing is the chance it gives to live a life that is fun and to move around the world with a group that shares the same views. I cannot say exactly where the tipping point for numbers of crew is that requires the captain to separate themselves, but if in doubt err towards maintaining a slight gap.

'Sir always eats last'

In naval parlance, 'sir' historically relates to officers, and 'chaps' refers to the sailors working for a living – no unisex terms were thought of since none were needed. As a junior officer, we often dined in among the sailors and as young men in our early 20s we entertained a hearty appetite. It seemed a tangential comment at the time and esoteric to leadership, but at one mealtime our training officer, a didactic navigator, observed the mountain of food I had placed on my plate and that I was at the head of the queue. He cautioned me, saying that true leadership comes from thinking of the needs of others and putting your own needs to the rear of the queue.

To this day I still enter the crew mess ten minutes late to ensure that if there are any shortages in the magnificent meal presented by the yacht chefs then it will be me who misses out. I know that 'Sir always eats last' is not at all about the crew mess meal and a lot more to do with the larger picture of putting those you lead first. At one point I was confronted by a colleague who, when taking the crew car keys (yet again) for a personal outing said, 'RHIP.' When I queried this, he expanded: 'Rank has its privileges.' My thought, which I kept to myself but seared into my memory, was 'Don't ever behave using RHIP as my guide!' Of course the message of this is not really about the crew mess buffet or a crew car, but these small examples help me set the tone for my bigger leadership goals.

'Don't run: it will scare the troops'

Not unlike eating last, this phrase harks back to a bygone time when colonial outposts built the wealth of nations – a time when officers wore safari suits and knee-high socks below baggy shorts. Saying 'don't run' is not only about feet moving at pace; it is a metaphor for approaching all problems at a physical, emotional and mental walking pace. When the pressure comes on, I can feel my speech speed up, my sentences shorten and my correspondence becoming curt. When I am flustered and my mind is racing, my leadership is heavily compromised and my team suffer. It is a struggle to confront each day composed and alert, not rushing wide-eyed to conclusions. My presence must be the calming influence, slowing events down and never 'running' in front of the crew.

If leadership were simple or not important then it would not be under constant discussion. Leadership is hard work and never ends. I recognised through military training the importance of establishing a leadership style that was authentic to me. I read the books and watched all the leaders I had exposure to, all the time trying to find my own self in there somewhere. Even with the reading and the watching I don't know if I have truly found myself yet and maybe only the lucky few do.

I do know that my experiences represent growth points and, as any weightlifter will tell you, to grow muscles you must tear, rebuild and tear again, and there will be some pain along this path. My leadership growth has involved many tears and repairs, and it has shaped who I have become along the way. As painful as many of these points have been, I would not have been able to move forwards without them.

As a footnote to military leadership reflection, I will finish with a story about a leader who was tested so far beyond the normal that his learning, as important and valid in all walks of life, is hard to visualise. Possibly more to the point, it is quite awful to visualise.

The Stockdale Paradox

The prisoner of war camp in Northern Vietnam was often referred to as the 'Hanoi Hilton', a cruel place where the horrors of a badly conceived war were lived. The ranking officer in the camp was Admiral Jim Stockdale. He survived, and on returning to normal life, wrote an incredible book, *In Love and War*. The book includes his recollections of time in the prison and the diary entries of his wife Sybil through the same period, noting that she was also living in an emotional prison created by the uncertainty of her husband's absence. I read the book, and highly recommend it, but it was in the retelling of a conversation between the best-selling author Jim Collins and Admiral Stockdale that his learning really hit home.

Collins asked the question everyone wanted to know the answer to: 'Who didn't make it out?' Stockdale's response was as surprising as it was enlightening: 'It was the optimists, those who kept on expecting the time in prison to end sooner, or for conditions to improve.' When their incarceration continued (Stockdale was imprisoned for six years) and conditions did not change, these optimists were crushed and their will to live was taken from them. The survivors were those who maintained their unwavering belief that at some point they would prevail, but until that moment came, they fully accepted their situation.

To draw a straight line between a North Vietnamese prisoner of war camp and a superyacht would, deservedly, arouse derision. Nevertheless, learning outcomes are transferable across all environments and what Admiral Stockdale lived through could be seen as a test far beyond what most of us are ever going to encounter in our modern and wonderfully privileged lives.

The quote that grabbed me in Collins' retelling of an incredible conversation with Stockdale was:

> You must never confuse the need on the one hand for absolute unwavering faith that you can and you will prevail in the end, with the need on the other hand, for the discipline to confront the most brutal facts as they are right now.

Read this again and maybe a third time. How many times as captain in a crew meeting had I thought softening a hard message about the programme, the owner's directions or some other change to the yacht that would have a direct impact on the crew was doing them a favour? I was lighting the torch of false hope – a dangerous beacon that would not illuminate the path for any crew member to successfully follow. Now, I recall Admiral Stockdale and feel that I am bound to tell the situation as it is, even if that is unpalatable. I give no false hope – a hope based on the unlikeliest events occurring in an equally unlikely sequence. This strategy happily helps me to remains true to my trinity of captaincy: communicate, clarify and motivate (see p. 105).

Critically, I enforce to the crew my belief in their ability to overcome whatever adversity we are facing as a team. If I didn't believe in them as individuals, as a team and my ability to lead them through, then there would be no chance of success.

188

Key takeaways

- Remember these four rules: 1) Learn your people; 2) Be friendly, but you are not their friend; 3) Sir always eats last and 4) Don't run: it will scare the troops.
- Don't trade in false hope and don't allow others to build optimism based on an unfounded belief.
- Motivate through being clear about what the terrain in front looks like. Say out loud just how steep the road is and that it is filled with pitfalls.
- Always finish with your belief in your team's ability to overcome the defined adversity.

Toolbox 3 – the Job

So far, I have demurred and said that while my views are that of a captain, the learning is transferrable. I am still going to hold this position, but in this toolbox, I am going to talk about boatie bits – dealing with shore parties and the driving part – the bit that is only 1 per cent of captaincy, but the 1 per cent that must be correct. As always, I hope there are transferrable learnings. At the very least, learning about these aspects of yachting may give you a chance to understand the perspective of another – and this alone is a great learning.

TAHITIAN BLUES

Sailing around the oceans is a challenge for all ships, but superyachts complicate this further as often the voyage is undertaken at short notice on the whim of an owner who prides themselves on bending the world daily to their preferences. For a yacht captain this translates into not quite knowing when and where you are sailing and who will be joining on arrival. To execute such a 'plan' requires a different set of communication skills than standing in front of the crew or writing to the office.

I recall the first time I set out to cross one of the major oceans and the obvious became my reality: there is no help in the middle. If something goes wrong, the captain must craft the solution. Yes, a captain has a crew, but the final decision must come from command. There seemed to be so many moving parts that at first it was hard to know where to focus. I am a weather watcher so this came naturally, but I learned the hard way I was not able to fully delegate the correct ordering of supplies, fuel and preparing the required clearance obligations that are unique to each destination. I needed to know for sure we would be OK.

I like to seek seafaring metaphors to explain life; sometimes they work and sometimes they don't. Crossing an ocean is an apt metaphor for many of life's challenges: you are alone at the point of decision, but there is help all around and there are many who have gone before me and many out there who are willing to help along the way.

The network that I have built to help me cross oceans and be ready on arrival is something I am proud and respectful of at the same time. I really have taken the time to nurture the relationships as, with no hyperbole, my life and that of the crew may depend on their support. In an earlier time in my career, I swaggered about thinking that as the captain of a great big yacht I was bringing business and should be received accordingly with due deference. Some years and oceans later I learned:

- I needed the support of local businesses more than they needed me.
- Carrying my humility was a lighter burden, and was received better, than carrying my ego.

With this awareness I was able to reframe my relationships with shore-based support companies and now cannot recall the number of times that showing this respect has meant the difference between success and failure. It was never the knowns that caught me; I always worked hard to ensure that if there was information available, I would find it. It was the 'unknown unknowns' that caught me.

One particular incident provides a good illustration of how important maintaining good shore relations can be to a successful superyacht ocean crossing. I was crossing the South Pacific, to prepare for the biggest family trip of the year for the yacht's owners and one day out from port, after an 18-day sea passage, I was informed there was a general strike that would restrict the supply of fuel and provisions. I let this sink in for a little while.

There was no plan B, and we had to make the visit work. I called the agent (we needed to speak more than emails would allow). He made it clear that while it was completely regrettable, there was nothing that could be done. Everything was closed.

I knew that if we could make it into the port then at least we would have a chance, whereas if we stayed at sea all would be lost. He agreed to see what he could do. I was aware that he had strong relationships and the yacht had been to the destination before, so my optimism bias had already kicked in. We were still 24 hours out from the port; surely something would fall in our favour. Late in the evening I received a message that the pilot was willing to board, and we could enter the port. There would be no line handlers on the dock, though, and the agent wanted to know if this was a problem. I immediately replied that this was no problem: we would send a boat ashore with our own crew to handle the yacht's lines and we would see the pilot the following morning.

I slept well that last night at sea now that the first stage in the plan was in place. The morning arrived, the pilot boarded, and the yacht berthed without incident on the southern side of the cruise pier we had used on previous visits. Our agent was waiting on the dock and greeted me with a warm smile and a gift bag of treats, as he always did. He said fuel was going to be the greatest problem, since the fuel terminal was on full strike and would not be as flexible as the tourist industries. His suggestion was that we fuel on completion of the trip. While this seemed a good idea, it was not possible. The yacht had sailed for an extended period in the Marshall Islands already and had arrived very close to reserve fuel levels.

I knew there would be a way, and I recalled how many years previously when delivering port services as an agent it would be very normal to 'grease the wheels of commerce'. It was never the amount that mattered, rather it was the way in which the 'coaxing' was delivered. I asked for a contact at the fuel facility and said that I would call them directly. This was not our normal approach, and I sensed the agent's hesitance. After all, I did not speak French, and the terminal staff would undoubtedly be

French speakers. To assuage the concern, I explained it was just to show the respect of making direct contact and that I hoped the strategy would work.

I rang the number and found to my surprise that the terminal manager spoke very credible English, liked Australia and was quite willing for me to visit. We agreed that I'd do so first thing the following morning. That evening in preparation I did three things: 1) I went for a run past the fuel facility to see if there was a picket line; 2) I had the chef bake a cake; and 3) I requested €5,000 from the purser.

I was driven to the port at eight the next morning and was met by two workers manning the gate. There hadn't been a picket the night before, so I guessed they manned the gate only at the peak times. I explained who I was and who I was coming to see. They were friendly and pointed my way for me to the upstairs office. The manager and I met and shared tea and a slice of cake. He invited two of the administration staff in, who also had a slice. I mentioned the friendliness of the two on the gate and, since they represented the entire terminal non-administrative workforce that day, two slices were also taken down to them.

Once pleasantries had been completed, I apologised and said I had to cut to the quick: we needed to fuel in the next three days. This was not a surprise to the terminal manager, but his response caught me off guard. He said that all the terminal staff wanted to work and were disappointed that they couldn't. It seems they were expected to stand down in solidarity with other sectors that had gone out on strike seeking better conditions. This explained the friendliness at the gate, and it joined the dots into a line that matched the feeling I had had when entering the terminal, which was well-presented with white rocks, neat tropical plants and clear, fresh signage. This was a professional operation with a team who took pride in what they did.

'What could I do to get fuel? Is there someone I could speak, negotiate with?' I asked. The manager said there was nobody I could negotiate with, but the symbolism of the yacht coming to the terminal would be good for him and his team. He was willing to speak with his own union representative to gain a one-day release to complete our works. I was amazed and appreciative of the offer, but I also knew waterfront unions are not the most understanding and that until we had the fuel hose connected then nothing had really been approved. I thanked the manager for his time, passed the workers on the gate and walked the 2km around the port back to the yacht. The town was noticeably quieter than usual, with many of the public sector workers not present.

By the time I reached the yacht I was perspiring heavily and looking forward to some water. I had underestimated the tropical heat, after too long spent in the yacht's air conditioning. My phone rang at precisely the moment my foot stepped on to the first step on the gangway. I saw it was a local number and was very surprised when the fuel terminal manager said, 'Captain, your ship can fuel, but you must be here in the next two hours and be gone by five today.' I was not ready with any other reply than, 'Yes, of course.' Then the terminal manager rang off.

We had to hustle to recall some crew from their day off, but it was not too difficult to work within the times the terminal manager had stipulated. We arrived, fuelled and were back at the cruise dock by 4.30 in the afternoon. We had needed to shorten the full order, but it did not matter: we were going to be OK for the trip. As usual, there was the agent standing on the dock, and our crew once again handled the berthing lines. I came straight down to see him, as the day's events seemed so peculiar. I stepped to the dock just as one of the inter-island ferries came back from Moorea. It passed quite close but did not impact our conversation. The ferry was of greater significance than I had known. Many of the senior and more militant union leaders had spent the day in conference in Moorea and were returning on the ferry. The fuel terminal and many other services on that day had tried to operate as discreetly as they could without the oversight of their union representatives.

It was a great day to have moved a 130-metre yacht across a port to fuel and back in 'stealth' mode. The cake was gone, but the €5,000 were returned to the purser. I never even thought to offer it as an incentive. In this situation, given the pride of the fuel depot workers, my offering would have been offensive; they just wanted to do their job. This story comes to mind, since although I have several times shown a financial appreciation for a great service, 'a tip' that was neither requested nor demanded, I have never once had to pay a commission (money that is expected and required to complete the work). I hear so often of 'needing to buy your way in' and it leads me to question the person saying this as if it were once the way. It no longer is. I think the greater success comes from showing respect to support companies who in turn will use their relationships and not the crassness of an envelope clumsily passed.

As a footnote: in the reverse situation I have only ever once in over 20 years been offered a commission from a supplier to use their service. The supplier and I knew each other well and had worked on projects before, so it was not uncomfortable, but I made it clear I wanted their support for the quality of their performance and not due to any chance of my own personal benefit. We joked through it and the supplier said it was just his way of showing gratitude for the business. The commission was 10 per cent, and I asked that this be shown on the invoice so the yacht owner could benefit. I do not share this as a chance to confirm my own integrity, rather as a reminder of how to handle this situation if it arises.

Key takeaways

- The relationships you build and maintain within yachting, especially with shore staff, are what will enable smooth sailing. This is particularly true when unexpected things happen, and they always happen.
- Be respectful of different cultures and try to speak to people in person if possible. Many will go out of their way to help you if you approach them in the right way.
- It is not worth losing your name and future career opportunities for a commission from a supplier – any supplier and of any value.

HOW NOT TO HANDLE A SUPERYACHT

When captaincy loomed on my career horizon all my thoughts turned to technical competence. I look back fondly to that time to just how naive I was. I fretted about learning engineering systems to the level required to make judgement decisions in an emergency, ship-handling concerns, ocean crossings, knowing who to speak with when purchasing for the yacht. It is some time since those early days and I look back on them with an 'if only' perspective. In the event, my greatest challenges were never technical. I had technical challenges, many of them, but they were not the events that shaped me or those that I still reflect on, observing layers of missed leadership opportunities.

If you skipped straight to this section at the expense of the 'soft' information about leadership and crew, then it is likely you are sitting with the naivety that I once had with regards to what it really takes to captain a superyacht. The fact is, job competence is the floor to your captaincy or any other role on board, not the height of where you want to take it. Of course, that is not to say ship-handling and competence are not important, though, as I learned one sweltering day in Darwin, Northern Australia.

My adventures in Darwin are decades in the rear-view mirror, though they shaped me professionally and I draw upon the technical skills often. I speak with pride of the learning I gained as a tug skipper and P&O Maritime area manager but rarely expose the truth that I had very little idea of how to fulfil either role or how I was racked daily with anxiety. I share the triumphs of the time but may neglect to share the tribulations, such as the first time I swept the tug's stern under the bow of a ship docked at Stokes Hill Wharf. The ship lowered a light line for the tug to tie the heavier towline to for them to pull up to their deck. Darwin is a tidal port and there was a current running. The tug needed to sit at least 30 degrees head-to this current to hold position. I was working hard on the controls just to maintain the position and the line had not yet been connected. I recall the momentary flash of awareness that I had no idea what was about to happen.

Imposter syndrome went from being a conceptual term (one not in common use in 1998) to a brutal reality as the tug took the load of the long line over the stern. The pivot point of the boat moved to the tow hook, the bow lifted and the tug bucked around like my daughter's horse Lio being ridden for the first time. My eyes widened and my teeth clenched. The tug was old, and the controls faced forwards, yet all cues came from looking aft, like backing a small trailer down a boat ramp. I was not intuitive with the controls, and I was doing a poor dance twisting backwards and forwards from looking at the ship to reacting on the controls.

I could see the deck rating, in his sun-faded orange overalls, having to hold firmly on to the ladder fixed to the side of the tug's cabin as I overcorrected. I also noticed his free hand was very close to the emergency handle for the wire that would release the trip-hook on the tow; I could see he had little confidence in the 'new guy' – me – in the wheelhouse.

I survived. The tug did heel uncomfortably but we stayed upright, and the deck rating did not need to trip the release. Over time and hundreds of moves my corrections became more subtle and I could 'feel' the ship and anticipate the requests from the pilots and the moves before they happened. Along with my beloved tug, I drove many other vessels during this time and their diversity became tools in my mariner's toolbox that I would take out in the future. One unique vessel in this group was the fuel barge.

The fuel barge was conceptually simple, yet in construction an overly complex and unreliable vessel. There had been many made over the years and the one that the company assigned to Darwin had been standing idle for some time – never a good thing. The fuel barge was among the Wallaby class of support craft. Wallaby boats were very capable small craft, 38 metres long, 10 metres wide and when filled drew 5 metres of water. Their true value was that they could carry over 1,000 tonnes of fuel and fresh water and deliver this at an exceptional pump rate of 200 tonnes an hour via a pipe of 20cm diameter. This was effectively 55 litres per second of fuel moving from the barge to the ship.

The simplicity/complexity paradox of the fuel barge extended to the propulsion. There was a 360-degree rotating 'leg' forward and a second aft and these were controlled by two rotating levers in the small wheelhouse. In description this is an effortlessly simple propulsion and manoeuvring solution. In execution it was anything but and the vessel was plagued by electronic control issues.

The challenges of the vessel did not hold us back: we would receive orders for all sorts of fuelling jobs and sail out into the harbour to complete them. Often the crew would complete a job with the tug, tie it to the barge, jump from the tug and then man the second vessel, dragging the tug alongside. We took pride in this lean approach to operating. Within all this working diversity, fuelling a submarine was the most challenging, since their small tanks meant the engineer would need to constantly adjust the pump rate and maintain communication with the submarine crew. The barge itself could not dock directly on the submarine; a special set of fenders were needed to keep the two craft separate.

With the United States Navy, the situation was even more tense. The tank dimensions and points we could connect to were restricted for their security needs, so we went in blind and followed the advice on arrival. We had just completed one of these fuellings and the pump engineer was gaining receipt signatures from the US crew. I did not interact with the crew on the submarine because I never left the wheelhouse during pumping. I always stood, eyes fixed to the hose, with the emergency pump stop button in reach. Fuelling submarines, the flow rate was greatly reduced, but still any delay in stopping could result in thousands of litres of fuel being spilled.

On this occasion all had gone well and we were making to depart. The cargo was 90 per cent of the vessel's weight and made a great difference to our load, so we had arrived heavy and were departing light. I am not sure how to best describe this beyond 'we arrived sticky and left slippery'. I waved to the deck crew to let the lines go and turned the two controllers to use the propulsion legs to clear the submarine. The legs work on an electrical signal to determine their amount of movement. When the voltage changes, the motor rotates until the voltage difference is reduced to zero. If the controller is turned past 180 degrees, the leg will follow the shortest path to return the voltage difference to zero.

This detail of the control logic was not known to me as we departed the submarine and even if it had been, the significance would probably have been lost. By increasing the power and then rotating the propulsion lever past 180 degrees, the forward leg rotated the 'wrong' way, and the front of the barge lurched ahead of the fender and collided with the submarine with a large ringing sound – think a hammer hitting an empty drum. The front of the barge went the wrong way as the leg rotated anticlockwise and not clockwise as I had wished for. I recovered and we were 20 metres clear of the submarine when I saw a group of American sailors coming on to their deck and looking around for the source of the noise. I smiled and waved and nothing more was heard of this little oops, the special relationship between the USA and Australia maintained for another day.

These events in combination with my own hypercritical self-assessment led to a growing awareness that the concept of 'learn by doing' may not be appropriate to captaincy. 'Fake it till you make it' might work well for a Silicon Valley start-up, but not so well at sea.

As a footnote to my time driving the tug: years later I would be on the other end of the towline as a marine pilot and these early experiences in Darwin gave me an enormous advantage of knowing what the tug can and cannot do. I was a good judge of a tug master and could feel quickly the difference between one who 'felt and anticipated' and one who was just hanging in there. All pilots can do this, and I realised that me not knowing what I was doing in Darwin had not been a secret; my lack of knowledge must have been apparent to all, even if they didn't let on. The same is true across all roles on board superyachts – there's no hiding incompetence or inexperience at sea.

KEY TAKEAWAYS

- Core competencies such as ship-handling are the entry-level requirements for any role, so you need to work to ensure you have the required level of proficiency – and preferably expertise – in these areas.
- If your workplace requires the use of equipment or repetitive processes, small or large, the only way to rise above entry-level competence is to master the tools to a level of unconscious competence. Your conscious thought needs to be kept free for when it goes wrong … and it will go wrong.
- The concept of 'learn by doing' is not appropriate to captaincy.

IMPROVING YOUR CORE COMPETENCIES

The terms 'coach' and 'mentor' are overused, seemingly interchangeable and not entirely consistent in their application. There is little by way of entry criteria to calling oneself a coach, which is why there are so many of them. This is not to belittle the importance of the role – more to highlight the need to carefully assess who is presenting themselves as a coach.

There is a view in seafaring that 'only another seafarer understands'. There could be some truth to this, so it's worth employing someone with relevant experience if you seek a coach to help you improve your skills. Remember, though, that you are not looking for a mirror, rather someone who can listen to you in a trusted environment and let you be the biggest component of the relationship. Next, you need to be willing to be honest with yourself and your coach, to allow them to listen and support your growth so that you can perform at the level you want and need to be at, as measured by your crew and the yacht owner.

I have been fortunate in having some wonderful mentors and role models. These were never formal relationships, which is a shame as it would have been better for both parties if there had been definition and accountability in the relationship. The one exception to this was during training and education as a marine pilot.

After weather and emergencies, ship-handling is a question I am always asked about. Not because it is the most important, but perhaps because it seems a black art to those not in the profession. Captains like to maintain this myth to enhance their status and, with autonomous shipping on the visible horizon, possibly to extend our use-by-date that little bit more. Prior to taking on the role of large yacht captaincy, I had fantastic exposure to boat- and ship-handling during a childhood spent driving small boats and racing yachts, undergoing naval training and even time spent on harbour tugs and offshore vessels. That all added up to nothing when I first stood alone on the bridge of a large yacht as we approached a harbour. The 1 per cent I could not get wrong just became my reality.

In the early days my sense of being the imposter was never more present than as I stepped on to the bridge ready to manoeuvre. As my ship-handling experience increased, this feeling reduced, but there was still something missing. It was not until I took a break from yachting that this missing piece was shown to me: I had not been creating adequate piloting briefings. Once I learned these lessons, I found they are universal; they extend beyond ship-handling and weaving them into my wider professional endeavours improved my performance, enjoyment and efficiency.

I had studied berth-to-berth planning, which is a fundamental part of modern navigation training. To non-mariners it may come as some surprise that historically ships planned to stop near the port and then the final approach was something that just 'happened'. There is context in that harbour entry was the realm of local pilots employed by the port to ensure the safety of the critical national asset, but nevertheless it was a clear shortcoming that ships approaching coasts around the globe never quite had a plan for how to reach their destination.

To address this shortcoming, training was placed into the international curriculum to ensure all ship's officers could competently plan to bring a ship to a berth. This was a good idea, but it was taught awkwardly. It focused on the how and not the why. It was a course in blocking out areas on a chart and making hypothetical secondary abort plans. I am not saying these are not important, but their context did not gel with me. It is always hard to focus on the how if the why is not clear.

I sought more information and read *Passage Planning Principles* – a book that was not hard to find, since it was compulsory and present on every bridge. The text made sense but still I was left wondering, what does the concept of passage planning truly mean?

Working with a mentor

It was only when I was training and working as a marine pilot that I finally understood the importance of defining and communicating the metrics of a manoeuvre and everything fell into place. 'What speed?' 'When?' and 'Why?' always need to be known, and this should be shared with everyone on the bridge. This was the answer to my question, 'Why is berth-to-berth planning so important?' It is always important because the entire bridge team must be able to assess every action of the manoeuvring captain/pilot against an earlier briefed and agreed plan. Any deviation from this plan is to be highlighted and the cause explained.

Communicating the plan in advance of its execution does not limit the captain; it allows the agreed plan to be varied in response to the conditions and circumstances at the time. Of even greater importance, it also allows a challenge from another member of the bridge team if the deviation cannot be validated or explained by the captain.

During my time training to be a marine pilot I was assigned several mentor pilots. I was intrigued that the more experienced pilots exercised more care and concern with their own pilotage briefings than the juniors. Their sketches of the ship's planned manoeuvre into port were precise, their briefing books contained photographs of landmarks relative to the pilotage and during execution they communicated to the entire bridge team what they were doing and why. They were using photographs and maplets to ensure that onboard captains and crews working in their second languages did not lose details, regardless of language; maritime is a visual environment and these tools helped remove ambiguity.

I had never seen this degree of planning before, nor thought to do this myself on yachts. Up to this point I would take control at some point on the approach to the port and while I would seek port information from the pilot, I set my own approach speeds and angles based on my best judgement at the time. By 'judgement' I mean that I was nervously going slowly without any cohesive plan. If I had been asked at any point what my plan was, I could not have answered.

My knowledge had advanced a step forward well before my training as a pilot when a cruise–ship-trained deck officer mentioned the 'ten through one method' of checking speed reductions against the ten cables of the last mile. This is a crude method, but in comparison to nothing, it provides a glide-path reference that can be amended with knowledge of a yacht's capabilities and then tailored to wind and tide. It is a little too cautious an approach for modern yachts, but it's useful as the first reference line that can be adjusted with knowledge.

Using this speed reference guide, I was now in a better place, but I still was not communicating my intentions to the pilot and the bridge team; I may as well have been there on my own. This changed when I got the chance to truly observe and be mentored by a professional ship-handler. New-entry pilots are assigned a mentor, a more senior pilot who inducts, trains and supports them in their struggle to move from former captain and nervous ship-handler to competent pilot. A competent pilot is expected to step on board any ship, at any hour, in any weather, take control and bring that ship safely to the port – a port whose continued operation is vital to the economic and even social well-being of the community, of the state. Since it happens out of sight of the wider population, until it goes wrong, nobody really cares.

In the first weeks before I was assigned my mentor I was nervous because there were some less-than-ideal role models in the pilot team. I did not want to model how to present myself as someone who saw themselves as a superior when they were a culturally ignorant visitor to someone else's workplace. Luckily, my assigned mentor pilot was more than I could have hoped for. Ian had emigrated from the UK to Australia and during 19 years as a Thames pilot he had learned from some of the best in the world and had encountered all the weather and tidal systems that the high latitudes of the UK could offer. He had then spent two years in a lower-paid and remote port once he made the move to Australia. Australia struggles to generate enough marine pilots to meet its needs, yet remains very restrictive to pilots seeking employment from overseas.

From my own selfish perspective, Ian's experience only made him a more compelling mentor. The North Queensland commercial port was a very different environment to those of my yachting experiences, my time in Darwin being possibly the only experience that came close. North Queensland has large tides, strong currents, steady wind and then strong gusts in the opposing direction. It was as unforgiving an environment as I could imagine. The ships were also very different: I'd gone from delightfully overpowered and easy-to-manoeuvre yachts to single screw bulk carriers that were not as reliable as the captains would seek to have you believe on boarding.

All of this my mentor took in his stride – in fact more than that, with thousands of ship moves already completed he expected everything to go wrong on every pilotage. He planned for it and while I would never be as bold as to call him out, I think he was at times disappointed when yet another ship departed or arrived seamlessly by his hand.

Sometimes the thing that is going wrong is barely discernible to the untrained eye. During one memorable departure of a fully loaded Cape Size (a Cape Size ship is 280 metres long, 52 metres wide and when loaded extends 18 metres below the water), Ian turned to me and said, 'See that?' I didn't, and he could tell that from my body language. He said to remind him to explain after the departure. He took the ship to the port limits safely, we both departed by helicopter and while walking back from the helipad after the four-minute flight I asked him what he had wanted to show me. He mentioned that at that point the bow of the ship was being pushed back by the water resistance as the ship 'cut' across the berth pocket.

This might need some explaining. The berth pocket was deeper than the departure channel and required the ship to move from a clearance of 5 metres under the keel to a clearance of just 90cm. Picturing this from below, the surface is a wall of water 280 metres long and 18 metres high that has to be relocated through a very small gap. There is a lot of hydrodynamics going on in this event, but let's just say the water did not appreciate being forced along this route and, unable to compress further, pushed back against the ship.

During the departure Ian had observed the ship being pushed back by this wall of water and that the bow was moving 0.2 knots in the wrong direction. This speed is almost imperceptible to the naked eye and Ian was using both the precise information from his portable pilotage navigation unit and a highly tuned sense of movement, gained from carrying out so many manoeuvres. While we were moving forwards, Ian therefore gave an engine astern order to move the pivot point to his advantage and the forward tug was increased to lift off at three-quarter power to recover the bow. The entire event was observed, acted upon and rectified within two minutes. Ian's point was if he had not acted at that point then the situation would have been very dangerous. With a smirk Ian asked, 'Do you know the fastest thing in the world?' I returned the smile and let the story play out. He continued, 'Brendan, the fastest thing in the world is a fully laden Cape Size bulk carrier moving half a knot in the wrong direction.'

It was a well-trodden joke, but Ian had a laugh all the same, then he steadied and followed through by pointing out how sharp a marine pilot's focus must be, even at the seemingly slow-moving moments. It was a great lesson and as the months progressed and as I moved from being an observer to being the pilot executing the pilotage, the lessons flowed. I never accepted anything Ian shared without asking follow-up questions. He warmed to this and my ship-handling education was accelerating to a point where I began to surprise myself with my ability to anticipate and react to seemingly unlinked events. When I eventually became a solo pilot and there was a complex move ahead, I would phone Ian before boarding the ship to communicate my plan and build my confidence. My first call on landing and clearing the helicopter's noise would also be to Ian to debrief. He was so good at his job that he could picture the move from the call.

Key skills

My mentor had given me tools to use that were centred around two aspects:

1. The plan.
2. The team available as a resource to execute the plan.

The planning began well before arriving at the ship. We would do the simple things of checking the radio battery was fully charged, but we went further and put a spare battery in our pockets 'just in case'. The portable pilotage unit never failed, but nevertheless we turned it on and calibrated it ashore, every time. We would visit the control tower to look at their weather information. Sure, we could look to reliable weather forecasting on our mobile phones, but the control tower had real data from wind sensors on the docks. We could also look to an array of cameras that would also show details like the wavelets on the water, which clarified the real-life impact of the wind and the tide. The actual ship movement plan included speed reference points (speed over ground), headings (over ground), abort points and a final docking plan. This is no different to the training I had received and not been able to put into use – it's just that it was my mentor pilot who stitched it all together and made it relevant.

Ian's training demanded I communicate my actions in real time during the pilotage and the manoeuvring. If at 5 cables to the berth the plan was to be at 5 knots and the ship was at 6 knots, I would now say, 'The ship is a knot above our agreed plan and I am comfortable with this, as it is countering the winds, but will reduce speed and report again as we pass 4 knots.' This narrative continued across all aspects of the plan and the ship's captain and bridge team were not only asked to respond but forced to engage. I now made sure the dialogue was two-way, with the crew's opinions being sought to the point of being demanded. This changed everything: I was no longer alone on the bridge; everyone was working with me. If a steward happened to be on the bridge and heard my narrative, they too should be able to monitor and be engaged. I was now leveraging all the people available to deliver a safe manoeuvre.

I was lucky to be appointed a great mentor, but this might not always be the case. I knew that if in my future there was no assigned mentor, I would need to identify the person I wanted to learn from and ask them to assist. I would talk to them and let them know I wanted to learn from them, and for every piece of information they provided, I would seek another two; I would be greedy.

My mentor had given me tools for pilotage, but more than this, they had provided a road map for life:

1. Test and verify equipment (or information).
2. Develop a plan, then communicate the plan.
3. Amend the plan in sympathy to the changing conditions and engage the input of others.
4. Support and challenge the plan.

KEY TAKEAWAYS

- Adequate piloting briefings are essential. They should include precise details of the ship's planned manoeuvre into port and include maplets and photographs of landmarks relative to pilotage, so that everybody, even those speaking a second language, clearly understand the task.
- This pilotage briefing must be communicated to the entire bridge team, with an explanation of the how and why. This way, the agreed plan can be varied in response to the conditions and circumstances at the time and another member of the bridge team can challenge the captain if a deviation cannot be validated. This communication should be two-way between the captain and the bridge team.
- You need to plan for everything to go wrong when it comes to pilotage.
- If no mentor is formally appointed, seek out a person you can learn from, and then make full use of the opportunity by asking lots of questions.
- Create a road map for how you will execute a task. Mine is:

  1. Test and verify equipment (or information).
  2. Develop a plan, then communicate the plan.
  3. Amend the plan in sympathy to the changing conditions and engage the input of others.
  4. Support and challenge the plan.

'Say the thing out loud'

These lessons are wonderful when applied to pilotage, but I find that they are equally applicable for my leadership endeavours. My ability to communicate and embrace the support of a team to safely bring a large ship into port gave me a new frame when I returned to lead in other environments. I would not say it was a silver bullet to success, but it certainly helped. It led me to introduce another phrase that I would hear myself using: 'Say the thing out loud.'

There is often a 'thing' that impacts you or those around you in a professional or personal environment. The thing in my yachting environment could be my fatigue, nervousness around new guests or a myriad of other factors. Saying what I can see as shaping my behaviour in the moment is not a free pass for rudeness but announces that my judgement is being affected by something – and this is not always negative. When I exercise, I find I have good ideas. When this happens, I introduce the notion to my team using the phrase, 'So this morning when I had a lot of oxygen in my brain, I thought of …' Doing so both acknowledges a fresh thought proposed in good faith and allows others to challenge it, since good ideas during a run may not always transfer into an executable plan.

FAILURE TO PLAN IS A PLAN TO FAIL

Expanding on the learnings of pilotage and not attempting to create a paint-by-numbers guide to ship-handling, there are some critical fundamentals that I think are missed in basic career training. The first involves a mantra that is spoken repeatedly in yachting: 'Go slow like a pro.'

I first heard this from one of my yacht deck officers one day and it made so much that I had been observing fall into place. He had been told, mentored even, into believing that docking as slowly as possible was the solution in all situations. I like to watch other yachts or ships entering port whenever I can, and I try to mentally pre-empt their manoeuvres as though I am on board with them. What confronted me when doing this was the fact that the yachts were slow, really slow, and there did not seem to be a structure to the approach. There might be an engine movement, a pause and then a correcting movement. What was happening?

I too had driven yachts like this – caution, caution and more caution. It felt sound: what could be wrong with going so slowly, since it allows time to correct the yacht's position? The answer to this is that 'go slow like a pro' does not work when there is 20 knots of beam wind or a 2-knot current astern. In these conditions the yacht captain must adjust the speed to avoid slipping sideways into a terrible situation.

So how to go from slow and cautious to piloting and docking at a safe speed and in control? It can be varied, but any variation must be announced. The speeds must be sympathetic to the weather and increased or decreased accordingly. What does this mean in practice? As I conducted the master–pilot exchange the narrative would go something like this:

At 1 mile from the dock the ship's speed will not exceed 10 knots. At 7 cables there is a large turn, and I will use this combined with a reduction in engine speed to achieve 5 knots at 5 cables. At this point I will be within 10 degrees of the approach heading and the tugs will be connecting. I will be assessing the effect of the current and the wind. I am expecting some beam wind and if this is the case we will hold speed longer. If not, I will be reducing from 4 to 3 knots by 3 cables. At 1.5 cables the ship's speed will not exceed 2 knots unless we are being set down. If this is the case I will have the aft tug prepare to layback astern as a safety option. If at any point you observe that I am not executing this plan, and I have not communicated why I am varying it, I expect you to challenge me.

A NOTE ON DISTANCE AND SPEED

A nautical mile is 2,000 yards.
A cable is 200 yards.
There are 10 cables in a nautical mile.
A knot is a measure of speed: 1 knot is 1 nautical mile per hour, or 1.852km/hr.

This is written as I would speak it: a monologue delivered at pace. The ship is still moving towards the berth, and I need to pass on all the information concisely and get on with the approach. The bridge team listening in their second language might miss some details, but this is OK. As we continue, I break down each stage in slower time as the event passes, prefacing it with, 'As I mentioned in the arrival brief...' In many ways saying this holds me accountable as much to myself as to the others on the bridge. I cannot say one thing out loud and do another without acknowledging the change to the others or to myself.

I learned this narrative from my mentor pilots and when I returned to yachts, I took it with me. I modified it a little to the environment, but what I found was that by committing to delivering a brief like this I controlled the yacht speed in a much more consistent way.

If you are an informed yacht reader then you may be shaking your head and saying, 'We have no limitation on engine movements, do not use tugs and rarely even have a pilot.' This is all correct, but it does not invalidate the concept of being able to brief and articulate the plan in real time. The reader who is working with a smaller yacht may still be shaking their head and saying, 'I am solo on the bridge as the deck officers are needed elsewhere.' If this is the case, and it often is on yachts, I recommend having an engineer or a member of the hotel team on the bridge as your accountability partner – they do not need to understand the manoeuvre to listen to your plan, watch you execute it and speak up when you deviate without announcement.

HOW TO CREATE AND DELIVER A ROLES AND RESPONSIBILITIES STATEMENT

Taking a good idea from senior marine pilots is an easy straight line, from one bridge to another. One idea that was not so direct, but equally as valuable, came from sharing an operating theatre (as an observer) with a cardiovascular surgeon who is also my good friend. Simon had invited me to join his team as he operated on an aortic aneurism. It was a truly amazing experience, and I did not expect to learn as much from it as I did.

Simon is an accomplished amateur guitarist with a weakness for 1980s big hair, soft ballad rock. As the attending surgeon the playlist was his to command. Guns N' Roses entered the room during prep and Bon Jovi seemed to own the first incisions. The screaming guitars were only paused a couple of times, the first by the taciturn theatre sister as everyone in the team, me included, introduced ourselves. It wasn't that the team did not know each other – Simon and the anaesthetist had worked together for 20 years and the theatre sister for 18. The purpose of the introduction was for everyone to state to the group their roles and responsibilities during the procedure. Simon's, for example, was: 'I am the attending surgeon, and I will be operating on an aortic aneurism, replacing the damaged aorta with a stent. The patient's name is xxx and I met with him this morning to confirm his understanding of this procedure.' The nurse then said, 'I am managing the instruments and will be keeping a track and trace of everything used. We will not close the chest cavity until I have confirmed the count of all equipment.' Even I had a speaking role: 'I am Simon's friend, here to observe only and will follow all instructions to not impede the work of others.' This last bit I was told was an important legal point that had to be made.

My recollections of these statements of roles and responsibilities do not do them justice. They were spoken with conviction and listened to with intensity. I was very impressed by this procedure and realised straight away how it could improve the performance on the bridge.

On returning to the yacht, I brought the bridge team together and shared my wonderful experience, learning outcome and idea to introduce the process into our workplace. It was met with some side glances, glazed eyes and a kind of 'OK, if you really want to' tone in the body language of the other bridge officers. Regardless of this mediocre response, the weight of my will prevailed and prior to the yacht's next port departure we all came together – captain, chief officer, the officers of the watch (OOW) and the fore and aft deck station bosses (bosun and OOW) – and introduced ourselves. I went first:

> Hello everyone. My name is Brendan. I am the captain and will have master's responsibility for all actions today. I will greet the pilot and conduct the master–pilot exchange and during this exchange I will make clear that while we take all advisement from the pilot, we will handle our own vessel. I will not be conning today; that will be role of the chief officer. If at any time I have a concern over the safety of the vessel to a point I need to take control I will do so by saying, 'I have the con.' If the conning officer at any stage feels a loss of awareness, he can ask me to take the con. At all times I retain responsibility.

The others followed my lead. We all felt uncomfortable stating the obvious to each other (and to the vessel data recorder) – it felt like theatre. I finished the pre-departure briefing thinking this might just be one of my 'wonderful ideas' that might not fly. After we had departed and once we were sailing in open seas, I asked the chief officer what he thought. He said it felt strange, but it focused him on what he was responsible for. With this as a slim endorsement I and the team carried on doing this in the future. Over time, it started to not seem strange and we no longer felt like third-rate actors trying to remember our lines.

Key skills

The roles and responsibilities statement, referred to as the 'Grey's Anatomy moment' in my future bridge teams, became a core part of our peak period safety protocol. I use the term 'peak period' as it was relevant at times other than just departures and arrivals. We often moved the yacht in remote areas where the chart data was scrappy to say the least and we also went through high-traffic areas. There was no script, but we did try to cover some key points each time. Use the list below as a starting point and modify it to your own environment. Don't be put off by how strange the first couple of times will feel for the team.

1. Who are you?
2. What are you doing today?
   a. Communications external/internal
   b. Operating equipment
   c. Leading others
   d. Authority levels/delegations
      (captain cannot delegate command)
   e. Concerns particular to the day
      (new roles/external factors)
3. Is this your normal role?
4. Is there a fallback position for your role today?*

*This last point is important. There are always two modes to any critical manoeuvre or transit: normal operations, and emergency. Often, bridge roles change between these modes and this needs to be addressed in the statement.

Have a think how such a structure can help your and your team's performance. Realise in advance that it will feel forced to begin with, but try to push through this phase. If you want to play with it, consider it a 'pre-mortem'. Talk about what can go wrong and how you can get out in front of this. In other industries this process might be called job safety analysis (JSA), but I find 'personal statement of responsibility' a better way to focus on the humans over the task when on board a superyacht.

As a final validation, if it were needed, just a few weeks ago (I am editing this text quite some time since it was first written) the 100-metre yacht I was captaining had a bow thruster failure 100 metres from the dock. It was a windy day and it was one of the most 'not good' failures that could have occurred. The bridge team, including the chief electrician, switched to secondary emergency roles and the fault was rectified without a voice or heartbeat being raised unnecessarily. The sentence was said, 'Moving to secondary emergency roles' and on assuming the con I was about to order both anchors to be dropped just as the electrician reported the machinery back in service. We then reverted to primary roles just as simply and the first officer took the con and docked the vessel. The cause? The second engineer had turned around next to the frequency converter for the bow thruster and his hip had hit the emergency stop. This was freakishly unlikely to occur, but such things happen.

KEY TAKEAWAYS

- By speaking roles and responsibilities out loud to each other (and to the vessel data recorder) you make yourselves accountable to each other.
- You will also have a chance to listen for gaps or overlaps in the plan's execution. In this format it becomes easy to challenge each other because it is not personal: it is the plan that is being challenged.

Guest anecdote: Da Gama Maritime – Steve Monk

I first met Steve Monk in 2009 during the construction of a yacht in Plymouth, UK. He was still in Royal Navy uniform and spoke of training and attention to safety in the yachting community. Since that time, I have watched him tirelessly campaign to improve standards within the industry. Once, I may have said he was a lone voice screaming into an empty room, but as expectations increase (and sadly) the frequency of accidents also increases, the room is filling. It is fun to read Steve's words; the naval officer within him remains strong and the tone is forceful – as it should be for the topic he is addressing below.

The above chapter sounds great for a vessel with more than one person on the bridge of the yacht during berthing manoeuvres and while that might be on the majority of vessels, I know there are many that are manned lean and thus operate in these most crucial situations with just the captain on the bridge. This is not ideal but reflects the reality.

With your [Brendan's] military and pilotage background, the closed loop and open communication methods on the bridge will seem more logical and obvious. It's the same way my father, being an advanced police Class 1 driver, taught me to drive by having to give a constant running commentary on what I saw, what I was doing and what I was anticipating. Going from that to the Royal Navy and navigating everything from minehunters to aircraft carriers made it natural to provide the running commentary on the bridge having given a fully detailed briefing to the captain beforehand.

The superyacht industry has snags but then so does the entire maritime industry, in so many ways that could be resolved simply. It all comes down to training, education and mindset. Too many people don't care or aren't interested. Their guide is, 'It's been fine for me for all these years so why should I change?' Obviously, your book is aimed at those who at least want to consider a change in attitude and ability so below is my contribution.

When considering a mentor or coach, ask yourself what is it you want out of the relationship and how is it going to be conducted? Utilising the apparent experiences of someone who doesn't work directly with you on the yacht should be considered as taking on a coach. They're not there to visually assess all your actions or to step in if it looks like immediate guidance is needed. They will have their advantages (if you choose well and are open to their suggestions) but ideally you may want to find a mentor – someone that works alongside you on the yacht right from the early stages of your career. Even as a deckhand or beginning steward(ess), you should be looking at those at the next rung on board to mentor you and enhance your knowledge.

As one proceeds up the ranks, your mentor will eventually be the captain who, if they have been guided well themselves, helps explain to you all the points from their experience which can be imparted, taught, explained, analysed and debriefed at the time.

Not all captains make great mentors and knowledge is often withheld, though as the industry matures, more will start to realise they can gain more professional satisfaction by helping others to progress and learn before mistakes are made.

Development opportunities may be found during berthing operations, coastal passage or open-water transits, but few bridge officers will be put under controlled stressful situations – ie simulators. Working through the career of most mariners, the time the average bridge officer spends in a simulator is considerably low, almost nil. So often the maritime industry is compared to the aviation industry and yet if that were the case, bridge officers would be recalled to professional simulator training on an at least annual basis to be observed, taught new skills and have pressure applied to controlled situations to witness reactions. Owners or guests may not be aware their 'new and fun' captain may have little or no experience handling a vessel of the size and propulsion of their new purchase. Ironically, at no time during standard training is a mariner assessed or signed off on being able to demonstrate driving a vessel. The poor individual, excited in the new role but with little driving *experience, will feel the weight of the world pressing down on them at these crucial times.*

To enhance the aspects Brendan has spoken about, the Royal Navy go further in their means of passage planning and inshore operations. Full briefings are provided on the intentions for departure or arrival, and these are briefed to the entire bridge team as well as the lead deck personnel of each station. With practice, these don't take long to prepare or give. Engineers will attend so that they and all others know what the plan is, what the secondary options are and what is anticipated based on weather, tide and other shipping movements.

Once on the bridge and physically executing the manoeuvre, while the captain may be driving the ship, the navigator will be looking wherever the captain isn't to ensure a full 360-degree coverage is maintained. The navigator provides a running commentary to ensure everyone on the bridge is aware of the intentions and reasoning for the actions about to take place. Experience is grown by 'feeling' what the ship is doing. Look out of the window to take into account every possible aid to navigation. Ashore transits, bearing drift of the bow and stern, feeling the delays to heading or speed changes as a result of shallowing water or closing in on underwater banks, looking at the flags on your own mast as well as those around you to get an indication of what the wind's doing, looking at the waves to get an indication of wind speeds and current direction are all basic factors that can help predict what's going to happen to the yacht as you progress.

Many yachts have electronic charting (ECDIS) slave displays on the bridge wing and subsequently bridge officers are becoming overly reliant on the information they provide. An ECDIS is a fabulous aid but cannot replace the fundamental sense of movement required for safe docking, gained from observation of relative movement. The line of the dock will provide the line of approach to best counter any wind or tide and subsequently what mooring line is your priority to hold the yacht and bring her alongside. You don't need an ECDIS for that, just eyes.

Any decent mentor or coach will tell you to plan and prepare: 'Don't let the first time something happens be the first time you thought about it.'

Those quiet times on watch or when planning your first departure should all be spent thinking about the what-ifs. You won't think of them all the first time you do this, but your experience will grow and as it does, you'll find yourself being able to mentor the junior officers and help them to understand what you're thinking and why.

All too often the phrase 'not enough time' is used in the superyacht industry when it comes to training or learning. This is wrong and demonstrates incorrect prioritisation. If you're looking to learn or want to be a good mentor, taking five minutes after a port departure or arrival to be able to discuss what went right, wrong or could be done better is crucial.

Looking at the wider opportunities to guide and mentor as the captain on board, when it comes to drills and exercises, don't micromanage. Don't try to do the jobs of your crew. You will have gone through the Advanced Firefighting course during which you will have been taught about the differences between strategic, tactical and operational actions. As captain, maintain your position of calm and authority by utilising the skills of your crew to feed the appropriate information up the chain of command to you, where you need only make the strategic decision you wish them to then action. Maintain the big picture and don't get sucked into the weeds.

Where the exercise reveals points for improvement in the debrief, work with the relevant crew to ensure they focus on those points but, more importantly, if they don't know specifically how to rectify them, reach out to external organisations and trainers or coaches for the assistance. No chief officer is expected to know every single safety procedure for every single emergency, particularly when these change as technology and lessons learned strive onwards. Working as a team on board to brief, debrief and enhance is all part of the demonstration of leadership in command and if you were never in that position or situation before, it should not be seen as a failure if you need to reach out to external coaches or trainers to seek advice and guidance. Prevention is better than cure and thus preparing for the worst is more effective and usually considerably cheaper than dealing with an unplanned emergency.

Never stop learning. Question others for ideas and options. Reach out to external companies for an outsider's look at what you do and how you do it. Good trainers aren't going to pull you apart and criticise what you do. They're there to review, guide, advise and, most importantly, help you to grow in confidence, capability, stature, standing and ability; to become a leader and someone others want to follow and aspire to be.

All ships are sailboats

Steve Monk's observations about ship-handling take me back to a time when I was interviewing for a pilot's position and one of the steps was a board-style interview. The board comprised of company representatives and two senior pilots in the company. Among the expected recruitment questions were several pilotage scenarios for me to respond to. The questions were generic in nature. I was not a pilot yet and had not seen the port, though anyone entering marine pilotage should be competent enough to go some way towards answering such questions.

The senior pilot began. 'You are on a bulk carrier in ballast. You are approaching the port to dock starboard side alongside. There is an above-forecast 20-knot wind 30 degrees on your port bow that, as you slow down, begins to set you aggressively towards the port. You have time to recover, but need to act now. What would you do? You have your tugs connected, port bow and port quarter, and all systems are in good working order.'

This is a lovely question. It is easy to visualise, completely generic to any vessel in any port and allows for several different solutions. The first that is apparent is to have the forward tug lift off at half to three-quarter power and have the stern tug (connected push/pull) to lift off one-quarter to stop the stern sliding as the bow lifts. There is a danger that in correcting the bow the stern will close the dock and there will be no recovery.

The second path is to do a similar manoeuvre with less power and thread the wind – make the wind one of your power sources and work its angles to your advantage. In parallel is to consider, pending proximity to dock, either more speed to provide longitudinal stability or if it was becoming dangerous, a short period with the engine astern to provide a bigger lever to the forward tug. In any case it would be worse to try and 'muscle' against the wind. I was quick enough to point out that the sail area of the hull was around 4,000 square metres and, even at a 20-degree angle, this was more tonnes of force from the 20 knots of wind than the tugs could overcome. I sensed this was the turning point in the interview and senior pilot Phil warmly said that he thought I would do OK. He finished this by sharing that 'all ships are sailboats'.

When I returned to yacht captaincy, I made this a refrain when deck officers were developing their skills. As they approached an anchorage or port, I would point out that the wind and current, when present, are the two most powerful thrusters available. If they are not harnessed, they will be working against you. In maritime, in life, the environment is your most dominant force, so do not ignore it.

HOW TO SET TWO ANCHORS

This is one toolbox skill that may breach the position I have been defending throughout this book: that all lessons are transferrable beyond superyachts and captaincy. That said, the 'power of having a plan' has been consistent and this story brings the theoretical to the actual, and this is transferrable learning even if you are never going to be watching a screen intently waiting for the perfect moment to release the second anchor.

One of my proudest maritime moments meant little to anyone aside from myself and a couple of bridge officers. The 10,000 gross tonne yacht was in Quiberon Bay, Brittany, France. We were bound there with a systems failure that prohibited passage to sea and while the engineers worked, I indulged my interest in extreme sailing and visited La Trinité-sur-Mer by tender to see the incredible 50ft and 60ft trimarans that the French sail so well. I knew the weather was forecast to deteriorate and this would be the last opportunity.

Filled with inspiration and photos I was returning to the yacht when I noticed the yawing had increased with the breeze, now gusting to 20 knots. I was not worried about the location of the yacht or the holding of the anchor, since the wind was from the north-west and the yacht was anchored as far to the north of the bay as was allowed within the harbour limitations. I was concerned with the yawing. This creates a risk of damage, makes working on deck difficult and fundamentally upsets me professionally. I returned to the bridge, briefed the OOW and began executing a precision two-anchor mooring. I approach the manoeuvre as a trigonometry problem to be solved. It requires some features to be available in the yacht's electronic charts (ECDIS), though these are normally within all systems.

Returning on board, I executed the manoeuvre. The wind was increasing and mid-manoeuvre I noted 35 knots as the clouds began to darken. Within three hours the wind would be a solid 50 knots, gusting to 60 knots. Where my pride came in was that in these conditions the yacht was yawing a maximum of 5 degrees. The tender tied to the stern was resting comfortably, the helicopters were sheltered behind the hangar, and the crew on deck were moving around with a stable breeze. None of those benefitting from the two anchors holding tight really cared that I was so proud of the manoeuvre that I felt compelled to screen-shot the ECDIS and to document the process. Over time this initial write-up has evolved, and then when my great friend and professional mentor, Malcolm Jacotine, added his thoughts – it became a well-rounded guide that I hope may answer some questions to the mariner reader.

Key skills

The most important component in two-anchor setting is to make the wind work in the yacht's favour and not to fight against it.

The process varies between open or confined waters. Modification for confined waters can be made simply. The principles remain the same, but the wind cannot be optimised in the same manner. In the absence of this, dropping the windward anchor will assist control. The execution of precision two-anchoring is one of the most rewarding and simple manoeuvres and the straightforward process below makes it replicable.

The planning prior to setting multiple anchors is crucial. In keeping with a running theme in this book, a plan with numerical metrics is essential. The same process also applies to setting lines astern to the rocks/berth. The activity of setting two anchors presupposes a dominant force in a predictable direction. Variable (frontal) winds/tidal waters may not be suitable for the practice.

ECDIS preparation
1. Over-zoom one ECDIS (keep a second ECDIS at the correct scale for safe navigation).
2. Ensure the SOG vector is on.
3. Ensure the wind arrow is visible and the variable range marker (VRM)/anchor circle panel is open.

Key metrics
As part of the preparation to any anchorage the following should be known:

1. Depth at the anchoring points.
2. Which anchor chain is to be used.
3. What is the scope created by the anchor chain.
4. Distance from shore (if using two anchors for stern-to-rocks).

Scope

The scope is the amount of anchor chain, though for better precision the length is more accurately represented as the opposite side in a right-angled triangle as follows:

- Opposite – scope for VRM calculation
- Adjacent – water depth
- Hypotenuse – anchor chain

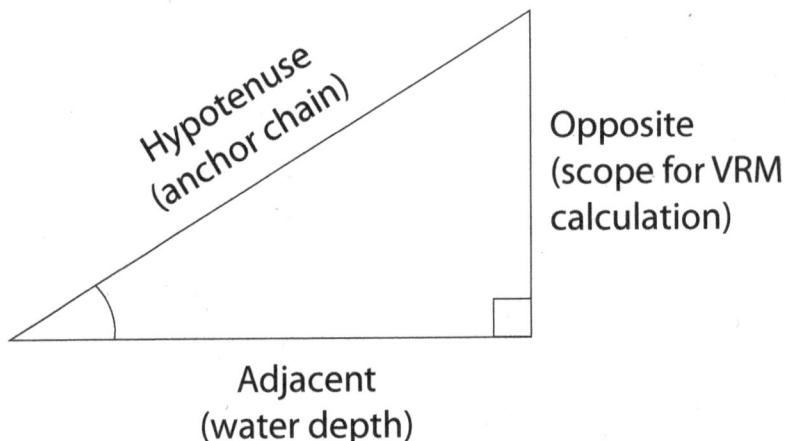

ANCHOR SCOPE

Anchorage selection
1. Identify the final vessel at anchor position (vessel – not anchor position).
2. Centre a floating VRM on this point (the range of VRM A + d in the diagram).
3. Identify wind heading and place the electronic bearing line (EBL) in the direction of the wind or current force.
4. Create a second VRM (50 per cent of the anchor scope) and centre this on the intersection of the first VRM and the EBL.

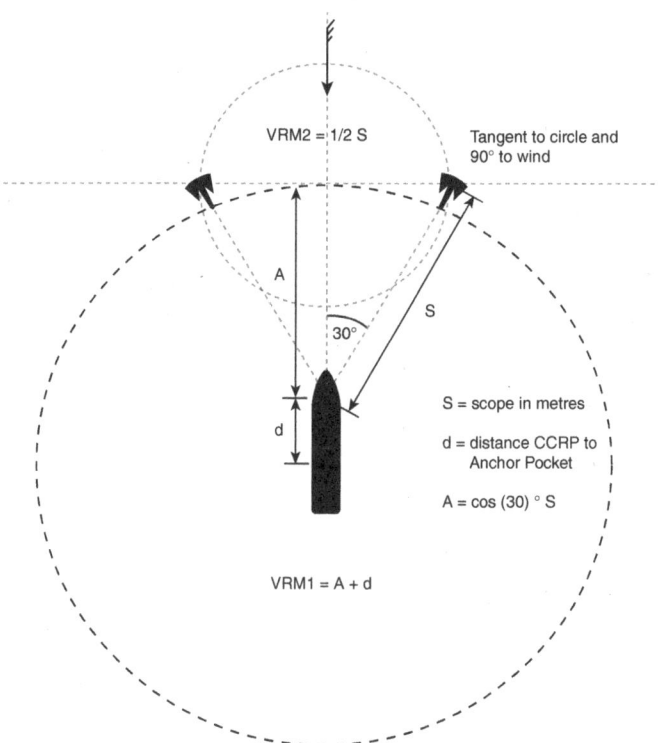

VRM2 = 1/2 S

Tangent to circle and 90° to wind

A

S

30°

S = scope in metres

d = distance CCRP to Anchor Pocket

A = cos (30) ° S

d

VRM1 = A + d

DIAGRAM COURTESY OF MALCOLM JACOTINE – WHO TOOK
A GOOD IDEA AND MADE IT BETTER

Execution

Bring the vessel to the first anchor position – using the wind on the side of the first anchor side (15–20 degrees). This allows the bow to be pushed away from the first anchor and the vessel to fall away towards the apex of the final triangle.

When in position:

1. Let go (or lower under power) the anchor.
2. Drop the position marker on anchor position 1.
3. Bring the bow through the wind so it is now sailing away from the first anchor.
4. Centre a VRM on the first anchor (radius 1/2 S) and move the EBL to be 90 degrees to the wind line.

** The set is to be 60 degrees apart (equilateral triangle), therefore the distance between anchors should be the same as the planned range of cable.

Second anchor

Pay out the first anchor as normal. The vessel should begin to track down the line that the cable will eventually lay (i.e. forming side one of the equilateral triangle) – this will vary depending on depth, but you should have (approximately) three to four shackles laid out before beginning to lift the vessel towards anchor position 2. The vessel is falling off the wind-line; don't worry about it – you are building a triangle.

Anchor 1 cable will be ranged an additional one to three shackles. Do not go short and try to drag anchor chain across the seabed – it will stop you from reaching anchor position 2.

The heading now is 20–30 degrees off the wind 'sailing' towards anchor position 2. Next:

1. When at anchor position 2, bring the vessel head-to-wind. It will naturally begin to move away from anchor position 2 (due to the weight of anchor 1).
2. Let go anchor 2.
3. Drop a second anchor mark (VRM) on ECDIS.
4. Keep the wind fine on anchor 2 bow to track down the anchor 2 VRM (second side of the isosceles triangle).
5. Simultaneously recover the excess cable from anchor 1 that allowed you to reach anchor 2 position. Do this gently and it will pull the ship towards the desired anchor position.
6. Utilise engines to build a slight sternway (approximately –0.7 knots) to stretch the two anchor cables.
7. Wait for the foredeck to confirm both anchors have been brought up.
8. Make final adjustments so that the vessel lies head-to-wind with both anchors at medium stay.

When two anchors are down, maintain watch on the vessel's movement using the track history. If the external forces (wind/current) change, the vessel will move to one of the anchors (leeward). When this occurs, shorten the leeward anchor. This may stabilise the vessel's movement, but if at any point there is a risk of turning on the anchors, just pick up the one at risk and return to one anchor.

Key takeaways

- The above method is not the only way to set two anchors and may not even be the best. Make it your starting point and adapt your technique to your own learning and preferences.
- 'Doing it by eye' is not enough on its own. There must be a plan and the execution should be treated with precision.

Toolbox 4 – Final Dos and Don'ts

It is all well and good for me to share my own experiences and treat them as some form of a truth. To really gain a perspective on just what makes a good captain or leader great, or a bad captain or leader terrible, I asked some colleagues past and present to 'give it to me straight'. When I looked at what came back to me, I saw some comments that I could recognise, for better or worse, and some others that were drawn from quadrants of their experience where our shared time did not overlap. In all cases, I found them valid.

Dos and Don'ts – A Stewardess' Perspective

Profile: Hotel-trained European stewardess, concierge, purser. A hospitality professional first, and a crew member second. Not shy of speaking out and wants to work alongside only professionals engaged at the same level she is.

Do:

- Maintain a good overview over what's happening in every department.
- Learn your crew and know what's happening with individuals on board.
- Make the effort to get to know your crew – sit with them in the crew mess during lunch/dinner, join in the social events every now and then, have a chat in the corridor etc.
- Be honest with the crew and share information where possible – try to avoid rumours spreading by being transparent.

- Promote/encourage personal development.
- Be an example to the crew – share your past personal experiences to encourage others.
- Have consistent rules for every single crew member.
- Make sure to let the crew know that your door is always open and that they can talk to you about anything at any time.

Don't:

- Let your crew play off against each other.
- Be too laid-back when it comes to safety of the vessel and the crew.
- Have double standards.
- Punish the whole crew when one person makes a mistake.
- Give the crew the feeling you don't have time for them.
- Make promises you can't keep.
- Be condescending towards the ones who didn't study to become a seafarer – i.e. interior crew.

I think all of these are valid, but if I were to draw out one it would be 'consistent rules for all crew'. Yachts are not tethered to large structural processes for crew leadership, which is a double-edged sword to be sure. Playing (unconscious) favourites is a great way for your embedded biases to flow through and upset the stability of the team you are trying so hard to lead. Transparency of decision-making is critical and providing a crew handbook with all (or as many as possible) foreseeable events documented goes a long way to overcoming this. There is not a copy-and-paste solution for this; you will need to do some work to make it authentic to your environment.

DOS AND DON'TS – A CHIEF OFFICER'S PERSPECTIVE

Profile: High-achieving cruise-ship experience, changeover to yachting, deck officer. Thoroughly knowledgeable and thoroughly professional. Does not suffer poor practice and is always challenging himself to progress. His comments reflect a strong mariner's concern for best practice.

Do:

- Maintain basics of captaincy/seamanship despite inevitable pressures and distractions.
- Maintain perspective (e.g. guest experience vs safety) – question yourself at every opportunity – what do you want to be remembered for?
- Encourage deck officers (and all crew) from any background, 'yachtie' or not.
- Make the most of the up-to-date/varied knowledge of above officers coming in from other areas of industry.
- Set high standards, lead by them and expect them of your crew.
- Cultivate a good culture on board – safe, hard-working, honest, healthy and enjoyable – it will filter through to the guests' enjoyment.
- Be human. Invest time in making it the best experience for guests balanced with a friendly work environment for crew.
- Delegate where possible – share workload and promote experience.
- Nurture and mentor the chief officer, when appropriate, to act as immediate deputy in order to ...
- ... SLEEP more when possible!

Don't:

- Fall into the trap of saying 'That's yachting' in a way that compromises safety – times have changed.
- Flout hours of rest regulations.
- Allow any level/severity of double standard on board.
- Expect everyone to conform to every one of your quirks/routines/desires (except standing orders), all the time.

- Settle for substandard crew for too long – make the best use of probation periods to avoid headaches later. Yes, people deserve a chance, but three months should be enough to prove they deserve the five-star position.
- Forget how small the industry is – build a brand for yourself from day one of captaincy.
- Shy away from holding management companies accountable when needed.
- Use only one weather service.
- Say no to the boss until you really have to; offer an alternatives first.
- Crash (the boat, or your body/mind!).

This is a challenging list for me to read, as I can see my flaws exposed under a former colleague's sharp eye. I talk a good game, but when on board with guests/yacht owners I am very prone to pushing myself and all around me to a place where fatigue is the defining factor in decision-making and the thinnest veneer of a safety culture is maintained. To address known and dangerous shortcomings I have learned to say 'no' and do so in a way that the guests I am looking after appreciate and revel in it.

It has become a fun superpower and begins as the guests board and receive their safety induction. I describe yachting as 'the land of yes' where all crew, me included, want to say yes to guests' every request. If for any reason a crew member is saying 'no' – where no may be linked to safety, where no may be due to resource limitations or no is because there is a crew boundary that must be respected – they are doing so not because they want to, but because they have thought long and hard before saying the humble two-letter word.

FINAL WORDS

I opened this journey into the heart of superyachting with the mention of a carrier wave, a simple premise that I sought to deliver a message using the yachts and their stories to carry a signal. I hope I have achieved this for all readers – the reader who is already deeply embedded in the yachting industry, the next who is considering entering or the last who is never going to cross the passerelle but is interested in what goes on 'out there on the water'. With waves of social media and reality television presenting a thin, though shameless, slice of #yachtlife I also wanted to plant a flag to say there are thousands of professionals at sea and ashore striving daily to deliver luxury tourism within this wonderful sector of maritime.

'The Industry' is a fabulous place to build a career and a life and to join a global community. This fact may be the one component that has not been given the attention it deserves in my stories from HMY *Mary* through to setting anchors. The superyacht diaspora is as broad and eclectic as one could want. From deeply passionate stainless-steel craftsmen to Monaco yacht sales brokers in their Loro Piana loafers who would not know one end of a welder from the other, all are fabulous and it is a pleasure to walk among them and call many friends. Sitting atop (my biased view) of all these participants in the yachting adventure are the crew – those who cross oceans, clean the cabins and fix the engines. As their captain, I love them, am frustrated by them and marvel at them in equal amounts (daily). I have raised my daughters within this realm and due to this choice must be an advocate that it can work. After looking at many alternatives, this option is easy to advocate.

Saying the advocacy is easy is not to be mistaken as saying 'the career is easy'. Yachting demands a lot from those who seek to make a career within its ramparts. Your calendar is not your own, job certainty is something you left behind in your 'pre-yachting' life and even when you've achieved superyacht success, the struggle does not end. At some point, leaving the cycle of eternal summers and exotic destinations to a return to the domesticity of shore life will not be easy; trust me, I have tried and yet I am still at sea for over half the calendar year.

I have also not hidden the traps embedded in superyacht culture, and the astute reader would be wise to take heed of my gentle warnings. I have watched respected peers from afar and in close-up lose their way trying to emulate the lifestyles of the yacht owners only to wake up one morning with an emptiness as they realise they were never part of this life and never will be.

As mentioned several (well quite a few) times, this is my second book and while not leading the reader to the ubiquitous 'users who bought this often buy this' button on the purchasing platform of your choosing, there was one piece of text from *Superyacht Captain* that I was particularly proud of. The only disappointing detail to this is that it was not written by me. I finished the book with Roosevelt's 'Man in the Arena' speech. This was chosen from my personal perspective, where I saw myself as 'the man' who has tried, fallen and tried again. The text will again be used for this closing, not because I could not think of better, but more that it is now provided to encourage you to step into your arena. Maybe it is a superyacht, maybe it is not, but step in to your choice regardless. Never sit on the sidelines commentating as 'the cynic'.

With love to you all.

Your humble and obedient servant,

Brendan

It is better to stumble than to do nothing or to sit by and criticize those that are 'in the arena' he explained. 'The poorest way to face life is with a sneer.' It is a sign of weakness. 'To judge a man merely by success,' he said, 'is an abhorrent wrong.'

It is not the critic who counts; not the man who points out how the strong man stumbles, or where the doer of deeds could have done them better. The credit belongs to the man who is actually in the arena, whose face is marred by dust and sweat and blood; who strives valiantly; who errs, and comes short again and again, because there is no effort without error and shortcoming; but who does actually strive to do the deeds; who knows the great enthusiasms, the great devotions; who spends himself in a worthy cause; who at the best knows in the end the triumph of high achievement, and who at the worst, if he fails, at least fails while daring greatly, so that his place shall never be with those cold and timid souls who know neither victory nor defeat.

Theodore Roosevelt,
'Citizen in a Republic', 23 April 1910

FURTHER READING

This list contains many of the books that have informed and inspired my writing.

Brassey, Anne, *A Voyage in the 'Sunbeam', Our Home on the Ocean for Eleven Months* (London: Longmans, Green, and Co., 1878)

Brockman, John, *This Explains Everything: Deep, Beautiful, and Elegant Theories of How the World Works* (London: HarperCollins, 2016)

Cheney, Dorothy L. and Seyfarth, Robert M., *Baboon Metaphysics: The Evolution of a Social Mind* (Chicago: University of Chicago Press, 2007)

de Botton, Alain, *Status Anxiety* (London: Hamish Hamilton, 2004; New York: Penguin Books, 2005)

de Crespigny, Richard, *QF 32* (Sydney: Macmillan Australia, 2012)

Easterbrook, Gregg, *A Moment on the Earth: The Coming Age of Environmental Optimism* (New York: Viking, 1995; London: Penguin Books, 1996)

Easterbrook, Gregg, *The Progress Paradox: How Life Gets Better While People Feel Worse* (New York: Random House, 2003; London: Random House, 2004)

Gladwell, Malcolm, *Blink: The Power of Thinking Without Thinking* (New York: Little, Brown and Company, 2005)

Grant, Adam, *Think Again: The Power of Knowing What You Don't Know* (New York: Viking, 2021)

Hastings, Reed and Meyer, Erin, *No Rules Rules: Netflix and the Culture of Reinvention* (New York: Penguin Press, 2020)

House, David, *Passage Planning Principles* (London: Routledge, 2015)

International Maritime Organization. 'The International Safety Management (ISM) Code and guidelines on implementation of the ISM Code (London: IMO Publishing, 2018). Available at: www.imo.org/en/OurWork/HumanElement/Pages/ISMCode.aspx

O'Shannassy, Brendan, *Superyacht Captain: Life and Leadership in the World's Most Incredible Industry* (London: Adlard Coles, 2022)

Parrish, Shane, *The Great Mental Models Volume 1: General Thinking Concepts* (Ottawa: Lattice Publishing, 2019)

Pink, Daniel H., *Drive: The Surprising Truth About What Motivates Us* (New York: Riverhead Books, 2009; Revised edition, 2011)

Pressfield, Steven, *The War of Art: Break Through the Blocks and Win Your Inner Creative Battles* (New York: Black Irish Entertainment, 2002)

Rozwadowski, Helen, M., *Fathoming the Ocean: The Discovery and Exploration of the Deep Sea.* (Cambridge, MA: Harvard University Press, 2005)

Schmidt, Eric, Rosenberg, Jonathan and Eagle, Alan, *Trillion Dollar Coach: The Leadership Playbook of Silicon Valley's Bill Campbell* (New York: Harper Business, 2019)

Seuss, Dr., *Oh, The Places You'll Go!* (New York: Random House, 1990)

Slade, Rachel, *Into the Raging Sea: Thirty-Three Mariners, One Megastorm, and the Sinking of El Faro* (New York: Ecco/ HarperCollins, 2018)

Stockdale, Jim and Stockdale, Sybil, *In Love and War: The Story of a Family's Ordeal and Sacrifice During the Vietnam Years* (Naval Institute Press, 1990)

Syed, Matthew, *Bounce: The Myth of Talent and the Power of Practice* (New York: HarperCollins, 2011)

Walsh, Bill, *The Score Takes Care of Itself: My Philosophy of Leadership* (New York: Portfolio, 2009)

REFERENCES

'In 1846, Britain boasted 530 yachts': Bender, Mike, 'The Slow Expansion of Yachting in Britain, 1815–1870', *A New History of Yachting* (Woodbridge, Suffolk: Boydell Press, 2017) pp. 97–109

'It may have also helped the cause when the young Mr Darwin said in *Notebook M*': Darwin, Charles, *Notebook M*. Available at: www.darwin-online.org.uk/content/frameset?viewtype=side &itemID=CUL-DAR125.-&pageseq=1

'It has always seemed strange to me': Steinbeck, John, *Cannery Row* (New York: Penguin Books, 1994) p. 1

'Power distance indicates levels of authority': Shah, Rajiv, Gao, Zhijie and Mittal, Harini, *Innovation, Entrepreneurship, and the Economy in the US, China, and India: Historical Perspectives and Future Trends* (London: Academic Press: 2014)

'The beauty of the system': Herman, Edward and Chomsky, Noam, *Manufacturing Consent* (London: Vintage, 1995) p. 2

'Grant introduced the term "confident humility"': Grant, Adam, *Think Again: The Power of Knowing What You Don't Know* (New York: Viking, 2021) p. 46

'Implementing the safety and environmental protection policy of the Company': International Safety Management (ISM) Code. Available at: www.brainscape.com/flashcards/ism-general-12220453/packs /20989453

'Read at every wait; read at all hours', Cicero, Marcus Tullius (widely cited – not verified)

'Process saves us from the poverty of our intentions': King, Elizabeth. Quoted in Seth Godin, *The Practice: Shipping Creative Work* (London: Portfolio Penguin, 2020) p. 4

'the author Jim Collins speak of the "flywheel effect"': Collins, Jim, *Good to Great: Why Some Companies Make the Leap… and Others Don't* (New York: Random House Business, 2001)

'We assume a winning negotiation strategy is about talking': Fisher, Roger, Ury, William and Patton, Bruce, *Getting to Yes: Negotiating Agreement Without Giving In*. 2nd ed. (New York: Penguin Books, 2006)

'[We use] listening as a strategy of last resort' and 'I will pretend I am listening': Heen, Sheila. Interview by Shane Parrish, 'Negotiation Expert Sheila Heen: Decoding Difficult Conversations', *The Knowledge Project*, 30 April 2019. Available at: https://podcastnotes.org/knowledge-project/heen/

'Team psychological safety is defined as a shared belief': Edmondson, Amy, 'Psychological Safety and Learning Behavior in Work Teams', *Administrative Science Quarterly* 44(2) (1999): pp. 350–83. Available at: www.jstor.org/stable/2666999.

'Feedback must be given with positive intent': Hastings, Reed and Meyer, Erin, *No Rules Rules: Netflix and the Culture of Reinvention* (New York: Penguin Press, 2020)

'After hearing him explain his humble decision matrix': Schawbel, Dan, 'Neil Pasricha: Why It Is Possible To Achieve Work-Life Balance', *Forbes*, 11 April 2016. Available at: www.forbes.com/sites/danschawbel/2016/04/11/neil-pasricha-why-it-is-possible-to-achieve-work-life-balance/

'Well, the first rule is that you can't really know anything': Munger, Charles T., *Poor Charlie's Almanack: The Wit and Wisdom of Charles T. Munger* (Virginia Beach, VA: Donning Company Publishers, 2005)

'You must never confuse the need': Collins, Jim, *Good to Great: Why Some Companies Make the Leap… and Others Don't* (New York: Random House Business, 2001)

Roosevelt's 'Man in the Arena' speech: Roosevelt, Theodore, from the speech 'Citizenship In A Republic', 23 April 1910. Available at: www.worldfuturefund.org/Documents/maninarena.htm